About the Authors

Ron Larson, Ph.D., is well known as the lead author of a comprehensive program for mathematics that spans school mathematics and college courses. He holds the distinction of Professor Emeritus from Penn State Erie, The Behrend College, where he taught for nearly 40 years. He received his Ph.D. in mathematics from the University of Colorado. Dr. Larson's numerous professional activities keep him actively involved in the mathematics education community and allow him to fully understand the needs of students, teachers, supervisors, and administrators.

Ron Larson

Laurie Boswell, Ed.D., is the former Head of School at Riverside School in Lyndonville, Vermont. In addition to textbook authoring, she provides mathematics consulting and embedded coaching sessions. Dr. Boswell received her Ed.D. from the University of Vermont in 2010. She is a recipient of the Presidential Award for Excellence in Mathematics Teaching and is a Tandy Technology Scholar. Laurie has taught math to students at all levels, elementary through college. In addition, Laurie has served on the NCTM Board of Directors and as a Regional Director for NCSM. Along with Ron, Laurie has co-authored numerous math programs and has become a popular national speaker.

Laurie Boswell

Dr. Ron Larson and Dr. Laurie Boswell began writing together in 1992. Since that time, they have authored over four dozen textbooks. This successful collaboration allows for one voice from Kindergarten through Algebra 2.

Contributors, Reviewers, and Research

Big Ideas Learning would like to express our gratitude to the mathematics education and instruction experts who served as our advisory panel, contributing specialists, and reviewers during the writing of *Big Ideas Math: Modeling Real Life*. Their input was an invaluable asset during the development of this program.

Contributing Specialists and Reviewers

- **Sophie Murphy**, Ph.D. Candidate, Melbourne School of Education, Melbourne, Australia
  Learning Targets and Success Criteria Specialist and Visible Learning Reviewer

- **Linda Hall**, Mathematics Educational Consultant, Edmond, OK
  Advisory Panel

- **Michael McDowell**, Ed.D., Superintendent, Ross, CA
  Project-Based Learning Specialist

- **Kelly Byrne**, Math Supervisor and Coordinator of Data Analysis, Downingtown, PA
  Advisory Panel

- **Jean Carwin**, Math Specialist/TOSA, Snohomish, WA
  Advisory Panel

- **Nancy Siddens**, Independent Language Teaching Consultant, Las Cruces, NM
  English Language Learner Specialist

- **Kristen Karbon**, Curriculum and Assessment Coordinator, Troy, MI
  Advisory Panel

- **Kery Obradovich**, K–8 Math/Science Coordinator, Northbrook, IL
  Advisory Panel

- **Jennifer Rollins**, Math Curriculum Content Specialist, Golden, CO
  Advisory Panel

- **Becky Walker**, Ph.D., School Improvement Services Director, Green Bay, WI
  Advisory Panel and Content Reviewer

- **Deborah Donovan**, Mathematics Consultant, Lexington, SC
  Content Reviewer

- **Tom Muchlinski**, Ph.D., Mathematics Consultant, Plymouth, MN
  Content Reviewer and Teaching Edition Contributor

- **Mary Goetz**, Elementary School Teacher, Troy, MI
  Content Reviewer

- **Nanci N. Smith**, Ph.D., International Curriculum and Instruction Consultant, Peoria, AZ
  Teaching Edition Contributor

- **Robyn Seifert-Decker**, Mathematics Consultant, Grand Haven, MI
  Teaching Edition Contributor

- **Bonnie Spence**, Mathematics Education Specialist, Missoula, MT
  Teaching Edition Contributor

- **Suzy Gagnon**, Adjunct Instructor, University of New Hampshire, Portsmouth, NH
  Teaching Edition Contributor

- **Art Johnson**, Ed.D., Professor of Mathematics Education, Warwick, RI
  Teaching Edition Contributor

- **Anthony Smith**, Ph.D., Associate Professor, Associate Dean, University of Washington Bothell, Seattle, WA
  Reading and Writing Reviewer

- **Brianna Raygor**, Music Teacher, Fridley, MN
  Music Reviewer

- **Nicole Dimich Vagle**, Educator, Author, and Consultant, Hopkins, MN
  Assessment Reviewer

- **Janet Graham**, District Math Specialist, Manassas, VA
  Response to Intervention and Differentiated Instruction Reviewer

- **Sharon Huber**, Director of Elementary Mathematics, Chesapeake, VA
  Universal Design for Learning Reviewer

Student Reviewers

- T.J. Morin
- Alayna Morin
- Ethan Bauer
- Emery Bauer
- Emma Gaeta
- Ryan Gaeta
- Benjamin SanFrotello
- Bailey SanFrotello
- Samantha Grygier
- Robert Grygier IV
- Jacob Grygier
- Jessica Urso
- Ike Patton
- Jake Lobaugh
- Adam Fried
- Caroline Naser
- Charlotte Naser

Research

Ron Larson and Laurie Boswell used the latest in educational research, along with the body of knowledge collected from expert mathematics instructors, to develop the *Modeling Real Life* series. The pedagogical approach used in this program follows the best practices outlined in the most prominent and widely accepted educational research, including:

- *Visible Learning*
  John Hattie © 2009
- *Visible Learning for Teachers*
  John Hattie © 2012
- *Visible Learning for Mathematics*
  John Hattie © 2017
- *Principles to Actions: Ensuring Mathematical Success for All*
  NCTM © 2014
- *Adding It Up: Helping Children Learn Mathematics*
  National Research Council © 2001
- *Mathematical Mindsets: Unleashing Students' Potential through Creative Math, Inspiring Messages and Innovative Teaching*
  Jo Boaler © 2015
- *What Works in Schools: Translating Research into Action*
  Robert Marzano © 2003
- *Classroom Instruction That Works: Research-Based Strategies for Increasing Student Achievement*
  Marzano, Pickering, and Pollock © 2001
- *Principles and Standards for School Mathematics*
  NCTM © 2000
- *Rigorous PBL by Design: Three Shifts for Developing Confident and Competent Learners*
  Michael McDowell © 2017
- Common Core State Standards for Mathematics
  National Governors Association Center for Best Practices and Council of Chief State School Officers © 2010
- *Universal Design for Learning Guidelines*
  CAST © 2011
- Rigor/Relevance Framework®
  International Center for Leadership in Education
- *Understanding by Design*
  Grant Wiggins and Jay McTighe © 2005
- Achieve, ACT, and The College Board
- *Elementary and Middle School Mathematics: Teaching Developmentally*
  John A. Van de Walle and Karen S. Karp © 2015
- *Evaluating the Quality of Learning: The SOLO Taxonomy*
  John B. Biggs & Kevin F. Collis © 1982
- *Unlocking Formative Assessment: Practical Strategies for Enhancing Students' Learning in the Primary and Intermediate Classroom*
  Shirley Clarke, Helen Timperley, and John Hattie © 2004
- *Formative Assessment in the Secondary Classroom*
  Shirley Clarke © 2005
- *Improving Student Achievement: A Practical Guide to Assessment for Learning*
  Toni Glasson © 2009

Standards for Mathematical Practice

Make sense of problems and persevere in solving them.
- Multiple representations are presented to help students move from concrete to representative and into abstract thinking.
- *Think and Grow: Modeling Real Life* examples use problem-solving strategies, such as drawing a picture, circling knowns, and underlining unknowns. They also use a formal problem-solving plan: understand the problem, make a plan, and solve and check.

Reason abstractly and quantitatively.
- Visual problem solving models help students create a coherent representation of the problem.
- *Explore and Grows* allow students to investigate math to understand the reasoning behind the rules.
- Questions ask students to explain and justify their reasoning.

Construct viable arguments and critique the reasoning of others.
- *Explore and Grows* help students make conjectures and build a logical progression of statements to explore their conjecture.
- Exercises, such as *You Be The Teacher* and *Which One Doesn't Belong?*, provide students the opportunity to critique the reasoning of others.

Model with mathematics.
- Real-life situations are translated into pictures, diagrams, tables, equations, and graphs to help students analyze relations and to draw conclusions.
- Real-life problems are provided to help students learn to apply the mathematics that they are learning to everyday life.
- Real-life problems incorporate other disciplines to help students see that math is used across content areas.

Use appropriate tools strategically.
- Students can use a variety of hands-on manipulatives to solve problems throughout the program.
- A variety of tool papers, such as number lines and graph paper, are available as students consider how to approach a problem.

Attend to precision.
- Exercises encourage students to formulate consistent and appropriate reasoning.
- Cooperative learning opportunities support precise communication.

Look for and make use of structure.
- *Learning Targets* and *Success Criteria* at the start of each chapter and lesson help students understand what they are going to learn.
- *Explore and Grows* provide students the opportunity to see patterns and structure in mathematics.
- Real-life problems help students use the structure of mathematics to break down and solve more difficult problems.

Look for and express regularity in repeated reasoning.
- Opportunities are provided to help students make generalizations.
- Students are continually encouraged to check for reasonableness in their solutions.

Achieve the Core

Meeting Proficiency

As standards shift to prepare students for college and careers, the importance of focus, coherence, and rigor continues to grow.

FOCUS — *Big Ideas Math: Modeling Real Life* emphasizes a narrower and deeper curriculum, ensuring students spend their time on the major topics of each grade.

COHERENCE — The program was developed around coherent progressions from Kindergarten through eighth grade, guaranteeing students develop and progress their foundational skills through the grades while maintaining a strong focus on the major topics.

RIGOR — *Big Ideas Math: Modeling Real Life* uses a balance of procedural fluency, conceptual understanding, and real-life applications. Students develop conceptual understanding in every *Explore and Grow*, continue that development through the lesson while gaining procedural fluency during the *Think and Grow*, and then tie it all together with *Think and Grow: Modeling Real Life*. Every set of practice problems reflects this balance, giving students the rigorous practice they need to be college- and career-ready.

Major Topics in Grade 4

Operations and Algebraic Thinking
- Use the four operations with whole numbers to solve problems.

Number and Operations in Base Ten
- Generalize place value understanding for multi-digit whole numbers.
- Use place value understanding and properties of operations to perform multi-digit arithmetic.

Number and Operations—Fractions
- Extend understanding of fraction equivalence and ordering.
- Build fractions from unit fractions by applying and extending previous understandings of operations on whole numbers.
- Understand decimal notation for fractions, and compare decimal fractions.

Use the color-coded Table of Contents to determine where the major topics, supporting topics, and additional topics occur throughout the curriculum.

- ■ Major Topic
- ■ Supporting Topic
- ■ Additional Topic

1 Place Value Concepts

Vocabulary 2
- 1.1 Understand Place Value 3
- 1.2 Read and Write Multi-Digit Numbers 9
- 1.3 Compare Multi-Digit Numbers 15
- 1.4 Round Multi-Digit Numbers 21

Performance Task: Elevation Maps 27
Game: Place Value Plug In 28
Chapter Practice 29

2 Add and Subtract Multi-Digit Numbers

Vocabulary 32
- 2.1 Estimate Sums and Differences 33
- 2.2 Add Multi-Digit Numbers 39
- 2.3 Subtract Multi-Digit Numbers 45
- 2.4 Use Strategies to Add and Subtract 51
- 2.5 Problem Solving: Addition and Subtraction 57

Performance Task: Population 63
Game: Race to the Moon 64
Chapter Practice 65

- Major Topic
- Supporting Topic
- Additional Topic

viii

3 Multiply by One-Digit Numbers

	Vocabulary	68
■ 3.1	Understand Multiplicative Comparisons	69
■ 3.2	Multiply Tens, Hundreds, and Thousands	75
■ 3.3	Estimate Products by Rounding	81
■ 3.4	Use the Distributive Property to Multiply	87
■ 3.5	Use Expanded Form to Multiply	93
■ 3.6	Use Partial Products to Multiply	99
■ 3.7	Multiply Two-Digit Numbers by One-Digit Numbers	105
■ 3.8	Multiply Three- and Four-Digit Numbers by One-Digit Numbers	111
■ 3.9	Use Properties to Multiply	117
■ 3.10	Problem Solving: Multiplication	123
	Performance Task: Sound Waves	129
	Game: Multiplication Quest	130
	Chapter Practice	131
	Cumulative Practice	135
	STEAM Performance Task: Bridge	139

Multiplication Quest

Directions:
1. Players take turns rolling a die. Players solve problems on their boards to race the knights to their castles.
2. On your turn, solve the next multiplication problem in the row of your roll.
3. The first player to get a knight to a castle wins!

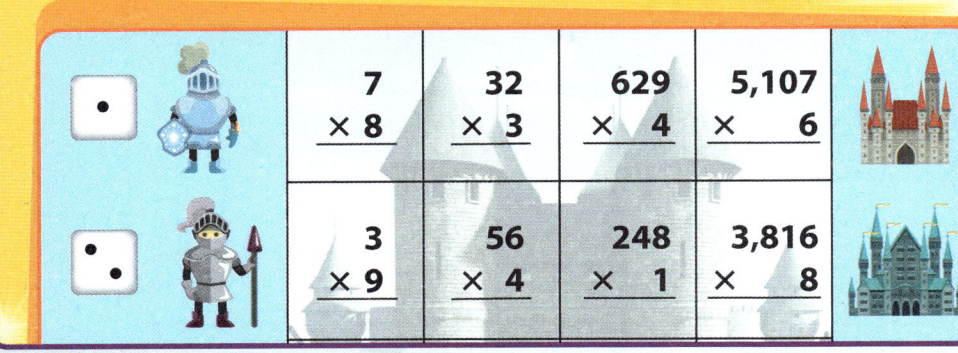

Row 1 (die shows 1):
7×8 32×3 629×4 $5{,}107 \times 6$

Row 2 (die shows 2):
3×9 56×4 248×1 $3{,}816 \times 8$

4 Multiply by Two-Digit Numbers

	Vocabulary	142
4.1	Multiply by Tens	143
4.2	Estimate Products	149
4.3	Use Area Models to Multiply Two-Digit Numbers	155
4.4	Use the Distributive Property to Multiply Two-Digit Numbers	161
4.5	Use Partial Products to Multiply Two-Digit Numbers	167
4.6	Multiply Two-Digit Numbers	173
4.7	Practice Multiplication Strategies	179
4.8	Problem Solving: Multiplication with Two-Digit Numbers	185

Performance Task: Wind Turbines ... 191
Game: Multiplication Boss ... 192
Chapter Practice ... 193

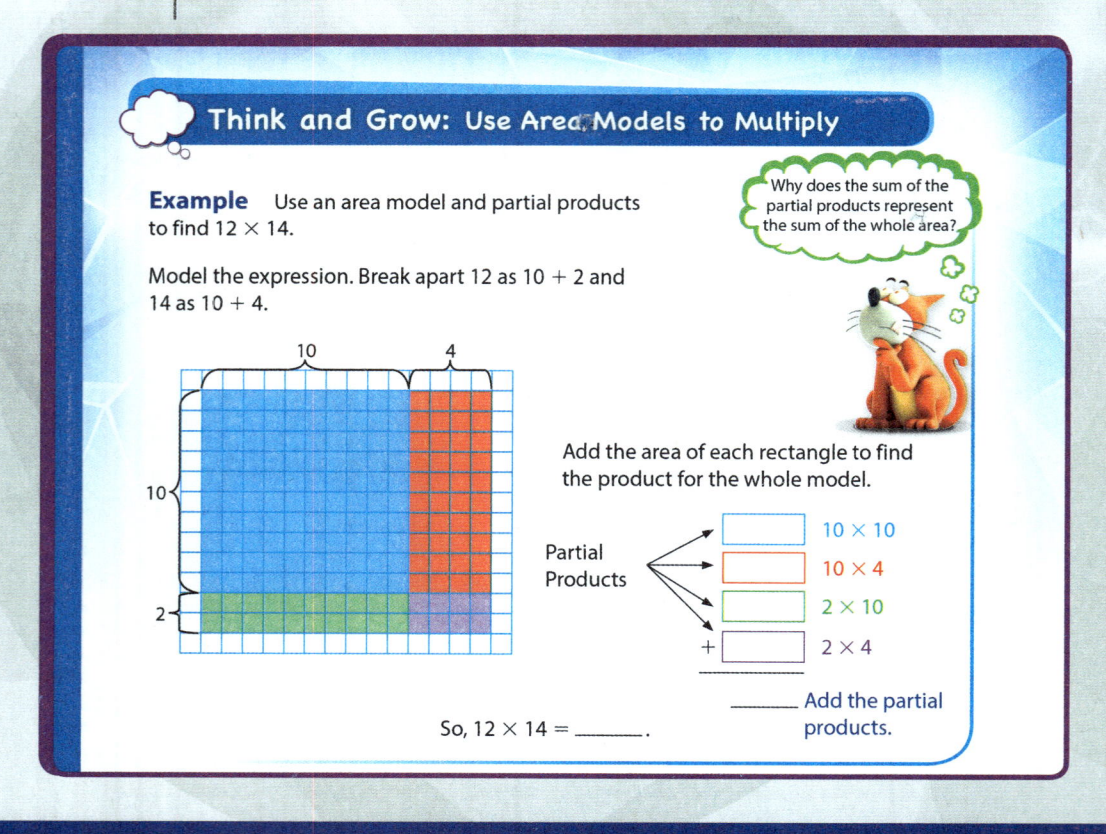

Think and Grow: Use Area Models to Multiply

Example Use an area model and partial products to find 12 × 14.

Model the expression. Break apart 12 as 10 + 2 and 14 as 10 + 4.

Why does the sum of the partial products represent the sum of the whole area?

Add the area of each rectangle to find the product for the whole model.

Partial Products:
- 10 × 10
- 10 × 4
- 2 × 10
- + 2 × 4

_____ Add the partial products.

So, 12 × 14 = _____.

x

5 Divide Multi-Digit Numbers by One-Digit Numbers

		Vocabulary	198
■	5.1	Divide Tens, Hundreds, and Thousands	199
■	5.2	Estimate Quotients	205
■	5.3	Understand Division and Remainders	211
■	5.4	Use Partial Quotients	217
■	5.5	Use Partial Quotients with a Remainder	223
■	5.6	Divide Two-Digit Numbers by One-Digit Numbers	229
■	5.7	Divide Multi-Digit Numbers by One-Digit Numbers	235
■	5.8	Divide by One-Digit Numbers	241
■	5.9	Problem Solving: Division	247
		Performance Task: Planetarium	253
		Game: Three in a Row: Division Dots	254
		Chapter Practice	255

6 Factors, Multiples, and Patterns

		Vocabulary	260
■	6.1	Understand Factors	261
■	6.2	Factors and Divisibility	267
■	6.3	Relate Factors and Multiples	273
■	6.4	Identify Prime and Composite Numbers	279
■	6.5	Number Patterns	285
■	6.6	Shape Patterns	291
		Performance Task: Basketball	297
		Game: Multiple Lineup	298
		Chapter Practice	299

■ Major Topic
■ Supporting Topic
■ Additional Topic

7 Understand Fraction Equivalence and Comparison

	Vocabulary	304
■ 7.1	Model Equivalent Fractions	305
■ 7.2	Generate Equivalent Fractions by Multiplying	311
■ 7.3	Generate Equivalent Fractions by Dividing	317
■ 7.4	Compare Fractions Using Benchmarks	323
■ 7.5	Compare Fractions	329
	Performance Task: Designs	335
	Game: Fraction Boss	336
	Chapter Practice	337
	Cumulative Practice	339
	STEAM Performance Task: Sea Level	343

8 Add and Subtract Fractions

	Vocabulary	346
■ 8.1	Use Models to Add Fractions	347
■ 8.2	Decompose Fractions	353
■ 8.3	Add Fractions with Like Denominators	359
■ 8.4	Use Models to Subtract Fractions	365
■ 8.5	Subtract Fractions with Like Denominators	371
■ 8.6	Model Fractions and Mixed Numbers	377
■ 8.7	Add Mixed Numbers	383
■ 8.8	Subtract Mixed Numbers	389
■ 8.9	Problem Solving: Fractions	395
	Performance Task: Music Notes	401
	Game: Three in a Row: Fraction Add or Subtract	402
	Chapter Practice	403

■ Major Topic
■ Supporting Topic
■ Additional Topic

9 Multiply Whole Numbers and Fractions

	Vocabulary	408
9.1	Understand Multiples of Unit Fractions	409
9.2	Understand Multiples of Fractions	415
9.3	Multiply Whole Numbers and Fractions	421
9.4	Multiply Whole Numbers and Mixed Numbers	427
9.5	Problem Solving: Fraction Operations	433
	Performance Task: Sounds	439
	Game: Three in a Row: Fraction Multiplication	440
	Chapter Practice	441

10 Relate Fractions and Decimals

	Vocabulary	444
10.1	Understand Tenths	445
10.2	Understand Hundredths	451
10.3	Fractions and Decimals	457
10.4	Compare Decimals	463
10.5	Add Decimal Fractions and Decimals	469
10.6	Fractions, Decimals, and Money	475
10.7	Operations with Money	481
	Performance Task: Baking	487
	Game: Decimal Boss	488
	Chapter Practice	489

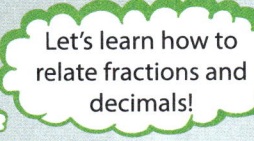

Let's learn how to relate fractions and decimals!

11 Understand Measurement Equivalence

	Vocabulary	494
11.1	Length in Metric Units	495
11.2	Mass and Capacity in Metric Units	501
11.3	Length in Customary Units	507
11.4	Weight in Customary Units	513
11.5	Capacity in Customary Units	519
11.6	Make and Interpret Line Plots	525
11.7	Units of Time	531
11.8	Problem Solving: Elapsed Time	537
11.9	Mixed Measures	543
	Performance Task: Racing	549
	Game: Conversion Flip and Find	550
	Chapter Practice	551
	Cumulative Practice	555
	STEAM Performance Task: Electricity	559

12 Use Perimeter and Area Formulas

	Vocabulary	562
12.1	Perimeter Formula for a Rectangle	563
12.2	Area Formula for a Rectangle	569
12.3	Find Unknown Measures	575
12.4	Problem Solving: Perimeter and Area	581
	Performance Task: Animations	587
	Game: Area Roll and Conquer	588
	Chapter Practice	589

- Major Topic
- Supporting Topic
- Additional Topic

13 Identify and Draw Lines and Angles

	Vocabulary	592
13.1	Points, Lines and Rays	593
13.2	Identify and Draw Angles	599
13.3	Identify Parallel and Perpendicular Lines	605
13.4	Understand Degrees	611
13.5	Find Angle Measures	617
13.6	Measure and Draw Angles	623
13.7	Add Angle Measures	629
13.8	Find Unknown Angle Measures	635
	Performance Task: Maps	641
	Game: Geometry Dots	642
	Chapter Practice	643

14 Identify Symmetry and Two-Dimensional Shapes

	Vocabulary	648
14.1	Line Symmetry	649
14.2	Draw Symmetric Shapes	655
14.3	Classify Triangles by Sides	661
14.4	Classify Triangles by Angles	667
14.5	Classify Quadrilaterals	673
	Performance Task: Roof Trusses	679
	Game: Pyramid Climb and Slide	680
	Chapter Practice	681
	Cumulative Practice	683
	STEAM Performance Task: Design	687
	Glossary	A1
	Index	A13
	Reference Sheet	A33

Let's learn how to identify symmetry!

1 Place Value Concepts

- What kinds of numbers would you find on a map?
- Why is place value important when you read a map?

Chapter Learning Target:
Understand place value.

Chapter Success Criteria:
- I can define the value of a number.
- I can explain how to use symbols to compare two numbers.
- I can compare the value of two identical digits in a number.
- I can read and write multi-digit numbers in multiple forms.

1 Vocabulary

Name _____

Organize It

Review Words
expanded form
standard form
word form

Use the review words to complete the graphic organizer.

- *word form* — three hundred fifty-seven
- *expanded form* — 300 + 50 + 7
- *standard form* — 357
- Ways to Write Numbers

Define It

Use your vocabulary cards to complete each definition.

1. period: Each group of __three__ digits separated by __commas__ in a multi-digit __number__

2. place value chart: A __chart__ that shows the __value__ of each digit in a __number__

3. thousands period: The period __after__ the ones period in a number

Chapter 1 Vocabulary Cards

| ones period | period |
| place value chart | thousands period |

Each group of three digits separated by commas in a multi-digit number

period			period		
Thousands Period			Ones Period		
Hundreds	Tens	Ones	Hundreds	Tens	Ones
1	0	0,	0	0	0

The first period in a number

Thousands Period			Ones Period		
Hundreds	Tens	Ones	Hundreds	Tens	Ones
8	1	5,	7	9	6

The period after the ones period in a number

Thousands Period			Ones Period		
Hundreds	Tens	Ones	Hundreds	Tens	Ones
8	1	5,	7	9	6

A chart that shows the value of each digit in a number

Thousands Period			Ones Period		
Hundreds	Tens	Ones	Hundreds	Tens	Ones
2	8	5,	7	4	3

Name _____

Round Multi-Digit Numbers   1.4

Learning Target: Use place value to round multi-digit numbers.

Success Criteria:
- I can explain which digit I use to round and why.
- I can round a multi-digit number to any place.

Explore and Grow

Write five numbers that round to 250 when rounded to the nearest ten.

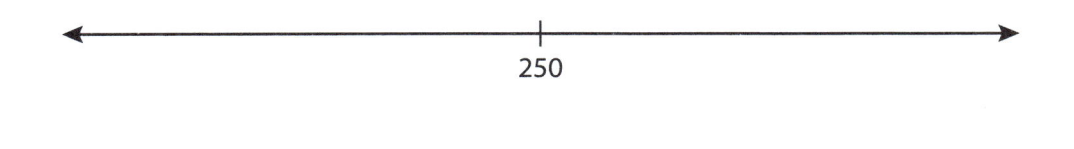

<---------|--------->
         250

_____   _____   _____   _____   _____

Write five numbers that round to 500 when rounded to the nearest hundred.

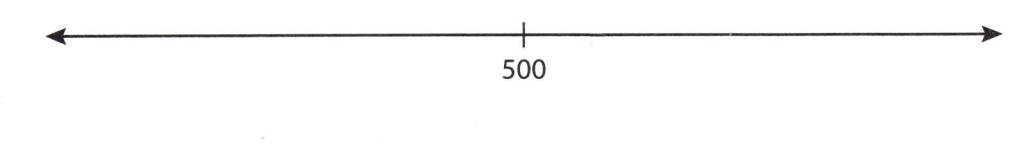

<---------|--------->
         500

_____   _____   _____   _____   _____

Repeated Reasoning Explain your strategy. Then explain how you could use the strategy to round a four-digit number to the nearest thousand.

Chapter 1 | Lesson 4   21

Think and Grow: Round Multi-Digit Numbers

To round a number, find the multiple of 10, 100, 1,000, and so on, that is closest to the number. You can use a number line or place value to round numbers.

Example Use a number line to round 4,276 to the nearest thousand.

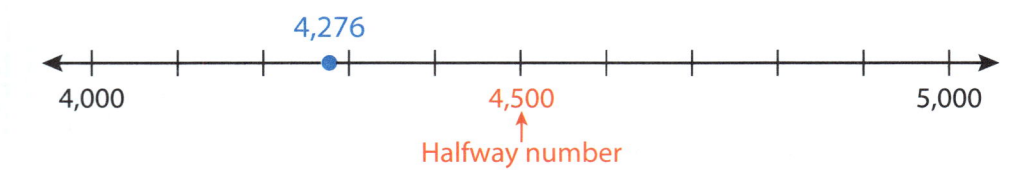

```
         4,276
←——+——+——+——●——+——+——+——+——+——+——→
 4,000            4,500                    5,000
                    ↑
              Halfway number
```

4,276 is closer to 4,000 than it is to 5,000.

So, 4,276 rounded to the nearest thousand is _____.

Example Use place value to round 385,617 to the nearest ten thousand.

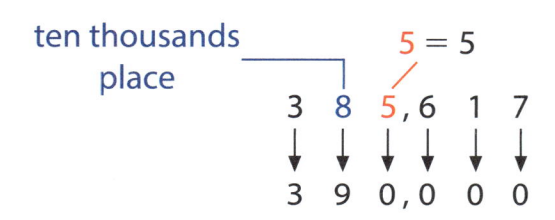

ten thousands place ———— 5 = 5

3 8 5,6 1 7
↓ ↓ ↓ ↓ ↓ ↓
3 9 0,0 0 0

> Remember, if the digit to the right of the rounding digit is 5 or greater, then the rounding digit increases by one.

So, 385,617 rounded to the nearest ten thousand is _____.

Show and Grow   I can do it!

Round the number to the place of the underlined digit.

1. <u>6</u>,912

2. 4<u>3</u>,215

3. <u>2</u>5,883

4. 1<u>4</u>8,796

5. Round 5,379 to the nearest thousand.

6. Round 70,628 to the nearest ten thousand.

7. Round 362,113 to the nearest hundred thousand.

8. Round 982,638 to the nearest thousand.

Name _____

Homework & Practice 1.4

Learning Target: Use place value to round multi-digit numbers.

Example Round 214,573 to the nearest hundred thousand.

Use place value.

hundred thousands place

1 < 5

2 1 4, 5 7 3
↓ ↓ ↓ ↓ ↓ ↓
2 0 0, 0 0 0

Remember, if the digit to the right of the rounding digit is less than 5, then the rounding digit stays the same.

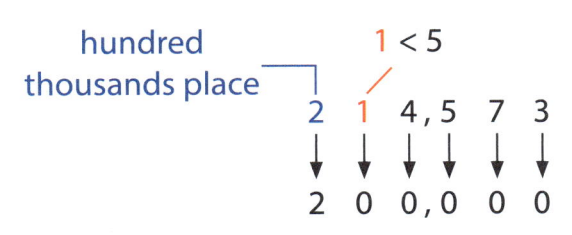

So, 214,573 rounded to the nearest hundred thousand is __200,000__.

Round the number to the place of the underlined digit.

1. 6,2<u>5</u>1

2. <u>8</u>,962

3. 3,<u>9</u>51

4. 7<u>9</u>,064

5. 43,9<u>7</u>6

6. <u>2</u>4,680

7. <u>7</u>26,174

8. 13<u>8</u>,691

Round the number to the nearest thousand.

9. 5,517

10. 70,628

Round the number to the nearest ten thousand.

11. 114,782

12. 253,490

Chapter 1 | Lesson 4   25

Round the number to the nearest hundred thousand.

13. 628,496

14. 90,312

15. **Structure** Round ★ to the nearest thousand and to the nearest ten thousand.

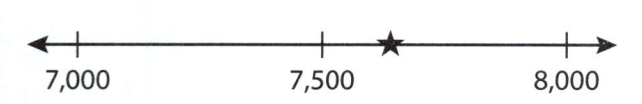

|←——+————+———★——+——→|
7,000   7,500   8,000

Nearest thousand: _____

Nearest ten thousand: _____

16. **Number Sense** Which numbers round to 300,000 when rounded to the nearest hundred thousand?

| 368,000 | 302,586 | 354,634 |

| 249,600 | 34,201 | 251,799 |

17. **Number Sense** When discussing the price of a laptop, should you round to the nearest thousand or the nearest ten? Explain.

18. **YOU BE THE TEACHER** Your friend says 5,953 rounds to 5,053 when rounded to the nearest hundred. Is your friend correct? Explain.

19. **Modeling Real Life** A glassblower is using a furnace to melt glass. When the furnace reaches about 2,000 degrees Fahrenheit, when rounded to the nearest hundred, she can put the glass in. At which temperatures could she put the glass into the furnace?

2005°   1899°   1925°   275°   2002°

Review & Refresh

Find the perimeter of the polygon.

20. 10 in., 4 in., 3 in., 6 in.

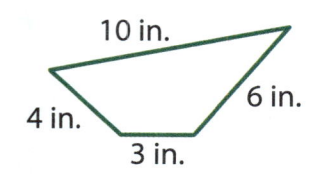

Perimeter: _____

21. Parallelogram
8 m, 5 m

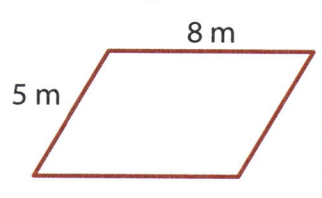

Perimeter: _____

22. Square
7 ft

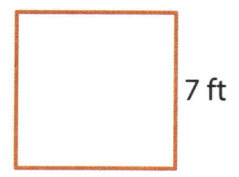

Perimeter: _____

26

Name _____

Performance Task 1

You hike from Point A through Point F along the orange path shown on the map.

1. What is the distance in elevation between each contour line?

   _____

2. As you walk from A to C, are you walking uphill or downhill? Explain.

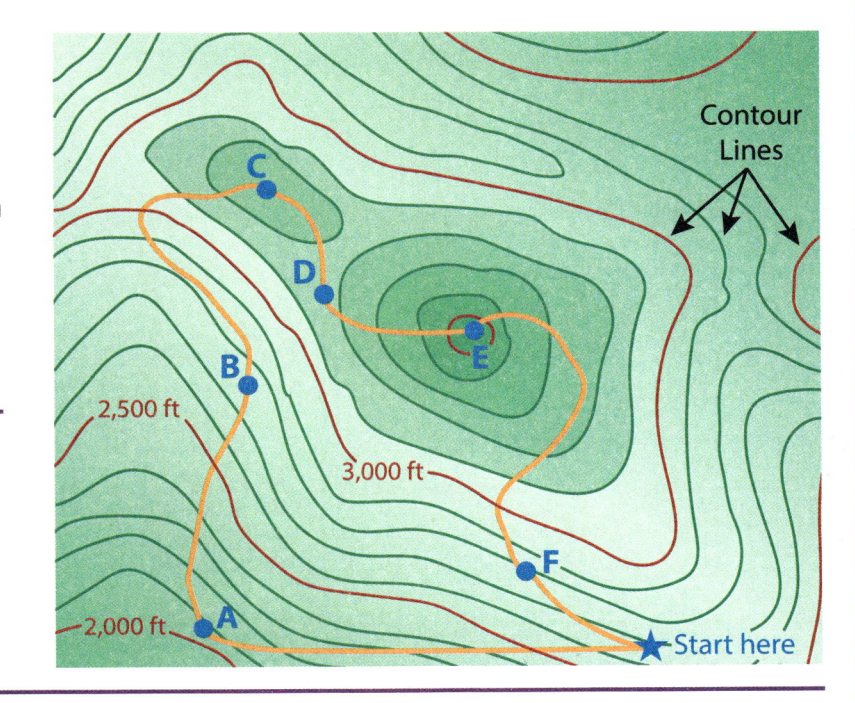

3. Which letter represents the highest point? Estimate the height.

   _____

4. A water station has an elevation of 2,763 feet. Which letter represents the location of the water station?

   _____

5. You take a break when you are at two thousand eighty feet. At which letter do you take a break?

   _____

6. About how much higher do you think Point C is than Point D? Explain.

Chapter 1   27

Place Value Plug In

Directions:
1. Players take turns.
2. On your turn, roll six dice. Arrange the dice into a six-digit number that matches one of the descriptions.
3. Write your number on the lines.
4. The first player to complete all of the numbers wins!

A number with...	Number
6 in the tens place 2 in the ten thousands place	___ ___ ___ , ___ ___ ___
4 in the ones place 5 in the thousands place	___ ___ ___ , ___ ___ ___
3 in the tens place 1 in the hundreds place	___ ___ ___ , ___ ___ ___
5 in the ones place 2 in the hundred thousands place	___ ___ ___ , ___ ___ ___
4 in the thousands place 6 in the ten thousands place	___ ___ ___ , ___ ___ ___
three of the same digit	___ ___ ___ , ___ ___ ___
digits in the hundred thousands place and ones place have a sum of 8	___ ___ ___ , ___ ___ ___
digits in the hundreds place and ten thousands place have a sum of 7	___ ___ ___ , ___ ___ ___
FREEBIE! Use any number!	___ ___ ___ , ___ ___ ___

Name _____

Chapter Practice 1

1.1 Understand Place Value

Write the value of the underlined digit.

1. 26,4<u>9</u>0
2. 5<u>7</u>,811
3. 30<u>8</u>,974
4. <u>7</u>48,612

Compare the values of the underlined digits.

5. <u>9</u>4 and <u>9</u>82
6. <u>8</u>17,953 and <u>8</u>4,006

1.2 Read and Write Multi-Digit Numbers

7. Complete the table.

Standard Form	Word Form	Expanded Form
		50,000 + 600 + 90 + 1
	seven hundred two thousand, five hundred	
993,004		

8. **Modeling Real Life** Use the number 10,000 + 4,000 + 300 + 90 + 9 to complete the check.

← standard form

← word form

Chapter 1
29

1.3 Compare Multi-Digit Numbers

Write which place to use when comparing the numbers.

9. 46,027
46,029

10. 548,003
545,003

11. 619,925
630,982

Compare.

12. 4,021 ◯ 4,210

13. 78,614 ◯ 78,614

14. 816,532 ◯ 816,332

15. 55,002 ◯ 65,002

16. 3,276 ◯ 3,275

17. 45,713 ◯ 457,130

18. 569,021 ◯ 50,000 + 6,000 + 900 + 2

19. thirty-seven thousand ◯ 307,000

20. Two different hot tubs cost $4,179 and $4,139. Which is the lesser price?

21. **Number Sense** Write all of the digits that make the number greater than 47,068 and less than 47,468.

47,___?___68

1.4 Round Multi-Digit Numbers

Round the number to the place of the underlined digit.

22. 8,6<u>1</u>4

23. <u>2</u>,725

24. 27,<u>6</u>02

25. 906,<u>1</u>54

Round the number to the nearest thousand.

26. 1,358

27. 57,094

Round the number to the nearest ten thousand.

28. 431,849

29. 60,995

30

2 Add and Subtract Multi-Digit Numbers

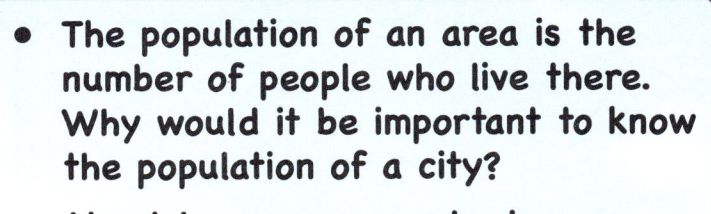

- The population of an area is the number of people who live there. Why would it be important to know the population of a city?
- About how many people do you think live in your city? Why is the population usually an estimate?

Chapter Learning Target:
Understand adding and subtracting numbers.

Chapter Success Criteria:
- I can estimate a sum or difference.
- I can explain which strategy I used to write a sum or difference.
- I can write a sum or difference.
- I can solve addition and subtraction problems.

2 Vocabulary

Name _____

Review Words
addends
Commutative Property of Addition
sum

Organize It

Use the review words to complete the graphic organizer.

[ ]

Changing the order of

[ ]

does not change the

[ ]

8 + 7 = 15   7 + 8 = 15

So, 8 + 7 = 7 + 8.

Define It

Use your vocabulary card to complete the definition.

1. estimate: A _____ that is _____ to an _____ number

32

Chapter 2 Vocabulary Cards

estimate

A number that is close to an exact number

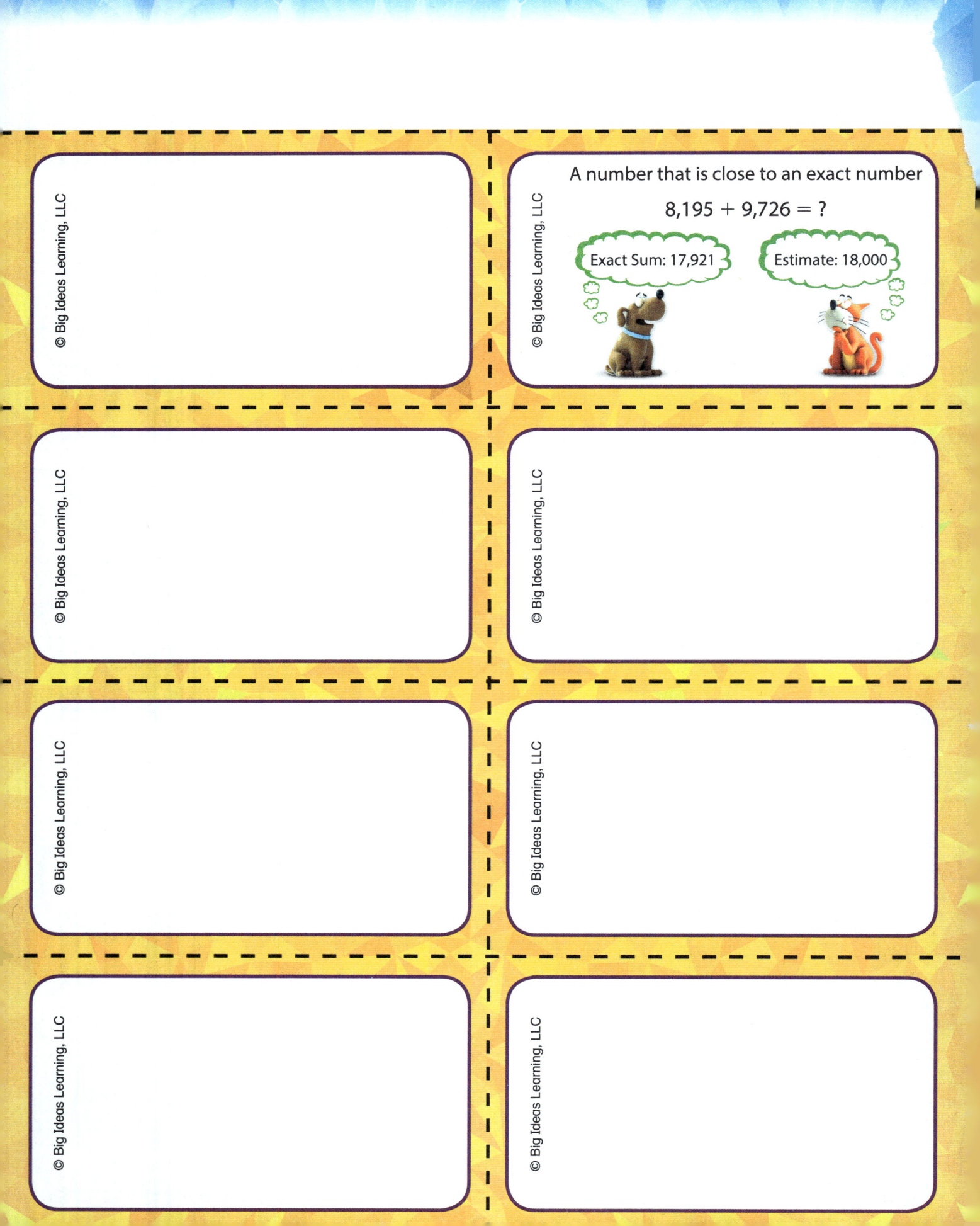

8,195 + 9,726 = ?

Exact Sum: 17,921

Estimate: 18,000

Name _____

Homework & Practice 2.1

Learning Target: Use rounding to estimate sums and differences.

Example Estimate 871,432 − 406,895.

One Way: Round each number to the nearest ten thousand. Then find the difference.

$$871,432 \rightarrow 870,000$$
$$-\ 406,895 \rightarrow -\ 410,000$$
$$\overline{}\ \overline{460,000}$$

So, 871,432 − 406,895 is about __460,000__ .

Another Way: Round each number to the nearest hundred thousand. Then find the difference.

$$871,432 \rightarrow 900,000$$
$$-\ 406,895 \rightarrow -\ 400,000$$
$$\overline{}\ \overline{500,000}$$

So, 871,432 − 406,895 is about __500,000__ .

Estimate the sum or difference.

1. 7,910 ⟶ ☐
   + 1,358 ⟶ + ☐
   ─────────
            ☐

2. 5,608 ⟶ ☐
   − 3,217 ⟶ − ☐
   ─────────
            ☐

3. 73,406
   − 45,699

4. 82,908
   + 28,643

5. 96,420
   − 63,877

6. 517,605
   + 359,421

7. 688,203
   − 444,387

8. 261,586
   + 116,934

Chapter 2 | Lesson 1

37

Estimate the sum or difference.

9. 864,733 − 399,608 = _____

10. 134,034 + 26,987 = _____

11. Number Sense Descartes estimates a difference by rounding each number to the nearest ten thousand. His estimate is 620,000. Which problems could he have estimated?

694,506 − 73,421

886,789 − 265,064

675,896 − 51,309

704,322 − 82,156

12. Reasoning When might you estimate the difference of 603,476 and 335,291 to the nearest hundred? to the nearest hundred thousand?

13. Modeling Real Life A storm causes 23,890 homes to be without power on the east side of a city and 18,370 homes to be without power on the west side. About how many homes altogether are without power?

14. Modeling Real Life You walk 5,682 steps. Your teacher walks 4,219 steps more than you. About how many steps does your teacher walk?

Review & Refresh

Find the product.

15. 3 × 3 × 2 = _____

16. 2 × 4 × 7 = _____

17. 3 × 3 × 5 = _____

18. 6 × 2 × 3 = _____

19. 4 × 9 × 2 = _____

20. 4 × 10 × 2 = _____

Name _____

Use Strategies to Add and Subtract — 2.4

Learning Target: Use strategies to add and subtract multi-digit numbers.

Success Criteria:
- I can use strategies to add multi-digit numbers.
- I can use strategies to subtract multi-digit numbers.

Explore and Grow

Choose any strategy to find 8,005 + 1,350.

Addition and Subtraction Strategies
Partial Sums
Compensation
Counting On
Regrouping

Choose any strategy to find 54,000 − 10,996.

MP Reasoning Explain why you chose your strategies. Compare your strategies to your partner's strategies. How are they the same or different?

Chapter 2 | Lesson 4

Think and Grow: Use Strategies to Add or Subtract

Example Add: 3,025 + 2,160.

One Way: Use partial sums to add.

$$3,025 = 3,000 + 20 + 5$$
$$+2,160 = 2,000 + 100 + 60$$
$$\overline{ 5,000 + 100 + 80 + 5 = \underline{}}$$

Another Way: Use compensation to add.

$$\begin{array}{r} 3,025 - 25 \\ +2,160 \end{array} \longrightarrow \begin{array}{r} 3,000 \\ +2,160 \\ \hline 5,160 \end{array}$$

You added 25 less than 3,025, so you must add 25 to the answer.

$$\begin{array}{r} 5,160 \\ +25 \\ \hline \end{array}$$

So, 3,025 + 2,160 = _____.

Example Subtract: 16,000 − 5,984.

One Way: Count on to subtract.

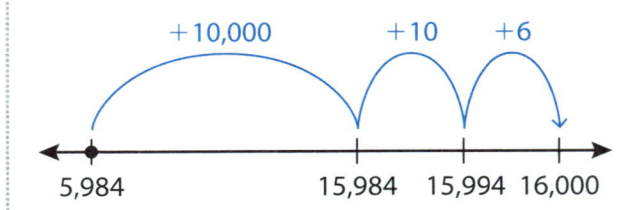

+10,000   +10   +6

5,984   15,984 15,994 16,000

10,000 + 10 + 6 = _____

Another Way: Use compensation to subtract.

$$\begin{array}{r} 16,000 \\ -5,984 + 16 \end{array} \longrightarrow \begin{array}{r} 16,000 \\ -6,000 \\ \hline 10,000 \end{array}$$

You subtracted 16 more than 5,984, so you must add 16 to the answer.

$$\begin{array}{r} 10,000 \\ +16 \\ \hline \end{array}$$

So, 16,000 − 5,984 = _____.

Show and Grow I can do it!

Find the sum or difference. Then check your answer.

1. $10,500$
   $+12,219$

2. $9,318$
   $-7,008$

52

Name _____

✓ Apply and Grow: Practice

Find the sum or difference. Then check your answer.

3. 12,982
  + 3,475

4. 44,561
  − 8,480

5. 91,803
  − 80,740

6. 87,871
  + 13,102

7. 246,132
  + 35,400

8. 860,709
  − 38,115

9. 780,649 − 13,754 = _____

10. 417,890 + 90,284 = _____

11. 614,008 + 283,192 = _____

12. 801,640 − 206,427 = _____

13. **Structure** Write an equation shown by the number line.

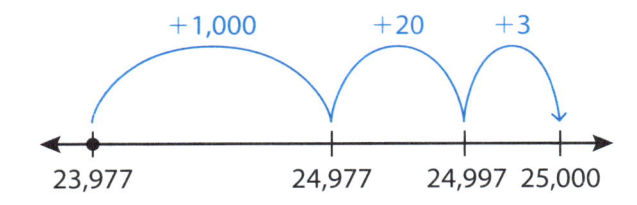

+1,000   +20   +3

23,977   24,977   24,997 25,000

Chapter 2 | **Lesson 4**   53

Think and Grow: Modeling Real Life

Example Earth's mantle is 1,802 miles thick. Earth's outer core is 1,367 miles thick. How many miles thinner is Earth's outer core than its mantle?

Subtract the thickness of the outer core from the thickness of the mantle. Use compensation to subtract.

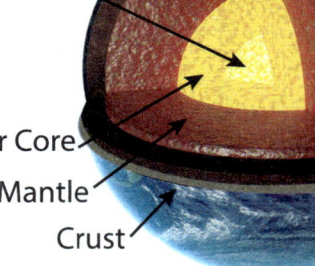

```
   1,802              1,802
 −1,367 + 33       − 1,400
                      402
```

You subtracted 33 more than 1,367, so you must add 33 to the answer.

```
   402
 +  33
 ─────
```

Earth's outer core is _____ miles thinner than its mantle.

Show and Grow   I can think deeper!

14. Your friend's heart beats 144,000 times in one day. Your heart beats 115,200 times in one day. How many fewer times does your heart beat than your friend's?

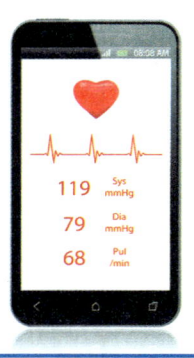

15. You, your friend, and your cousin are playing a video game. You score 2,118 more points than your friend. Your friend scores 1,503 fewer points than your cousin. What is each player's score? Who wins?

Player	Score
You	?
Friend	6,010
Cousin	?

16. Students at a school want to recycle a total of 50,000 cans and bottles. So far, the students recycled 40,118 cans and 9,863 bottles. Did the students reach their goal? If not, how many more cans and bottles need recycled?

Name _____

Homework & Practice 2.4

Learning Target: Use strategies to add and subtract multi-digit numbers.

Example Add: 4,165 + 2,730.

One Way: Count on to add.

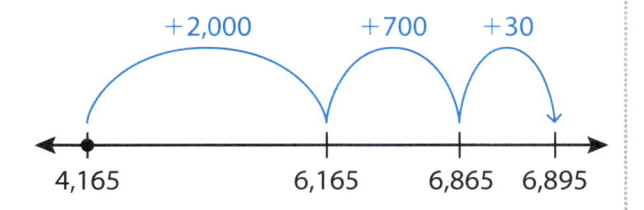

+2,000   +700   +30

4,165   6,165   6,865   6,895

So, 4,165 + 2,730 = __6,895__.

Example Subtract: 15,340 − 7,996.

One Way: Use compensation to subtract.

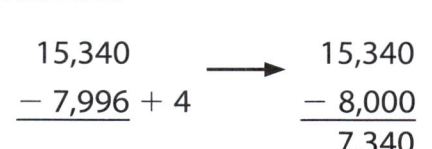

15,340   15,340
− 7,996 + 4   − 8,000
               7,340

You subtracted 4 more than 7,996, so you must add 4 to the answer.

7,340
+   4
7,344

So, 15,340 − 7,996 = __7,344__.

Find the sum or difference. Then check your answer.

1. 9,847
   + 3,907

2. 7,683
   − 4,653

3. 18,212
   + 7,589

4. 9,800
   − 2,175

5. 16,546
   − 8,031

6. 135,790
   + 26,145

7. 814,327 − 32,475 = _____

8. 294,801 + 46,030 = _____

9. 512,006 + 318,071 = _____

10. 746,620 − 529,706 = _____

Chapter 2 | Lesson 4   55

11. **YOU BE THE TEACHER** Your friend uses compensation to add. Is your friend correct? Explain.

    46,317 − 17   46,300   75,220
    + 28,920   →   + 28,920   −   17
                       75,220   75,203

12. **Writing** Which strategy would you use to subtract 9,618 from 58,007? Explain.

13. **Modeling Real Life** There are about 500,000 detectable earthquakes each year. About 99,500 of the detectable earthquakes are felt. How many earthquakes are detectable, but are *not* felt each year?

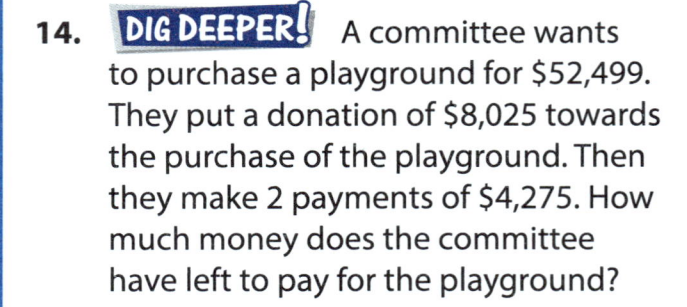

14. **DIG DEEPER!** A committee wants to purchase a playground for $52,499. They put a donation of $8,025 towards the purchase of the playground. Then they make 2 payments of $4,275. How much money does the committee have left to pay for the playground?

Review & Refresh

Write an equation to solve. Use a letter to represent the unknown number. Check whether your answer is reasonable.

15. There are 4 boxes. Each box has 6 granola bars. Your soccer team eats 18 granola bars. How many granola bars are left?

16. Newton has 35 blocks. Descartes has 28 blocks. Newton divides his blocks into 7 equal groups and gives Descartes one group. How many blocks does Descartes have now?

Name _____

Problem Solving: Addition and Subtraction 2.5

Learning Target: Use the problem-solving plan to solve two-step addition and subtraction word problems.

Success Criteria:
- I can understand a problem.
- I can make a plan to solve a problem using letters to represent the unknown numbers.
- I can solve a problem and check whether my answer is reasonable.

Explore and Grow

Use addition or subtraction to make a conclusion about the table.

| State Land Area ||
State	Area (square miles)
Arizona	113,594
California	155,779
Nevada	109,781
Utah	82,170

MP Precision Switch papers with your partner. Check your partner's answer for reasonableness.

Chapter 2 | Lesson 5   57

Think and Grow: Problem Solving: Addition and Subtraction

Example You have 3,914 songs in your music library. You download 1,326 more songs. Then you delete 587 songs. How many songs do you have now?

Understand the Problem

What do you know?
- You have 3,914 songs.
- You download 1,326 more.
- You delete 587 songs.

What do you need to find?
- You need to find how many songs you have now.

Make a Plan

How will you solve?
- Add 3,914 and 1,326 to find how many songs you have after downloading some songs.
- Then subtract 587 from the sum to find how many songs you have now.

Solve

Step 1:

3,914	1,326
k	

k is the unknown sum.

$3,914 + 1,326 = k$

$$\begin{array}{r} 3,914 \\ +\ 1,326 \\ \hline \end{array}$$

$k =$ _____

Step 2:

n	587
$k =$ ___	

n is the unknown difference.

_____ $- 587 = n$

$$\begin{array}{r} \\ -\ 587 \\ \hline \end{array}$$

$n =$ _____

You have _____ songs now.

Show and Grow *I can do it!*

1. Explain how you can check whether your answer above is reasonable.

Name _____

✓ Apply and Grow: Practice

Understand the problem. What do you know? What do you need to find? Explain.

2. There are about 12,762 known ant species. There are about 10,997 known grasshopper species. The total number of known ant, grasshopper, and spider species is 67,437. How many known spider species are there?

3. A quarterback threw for 66,111 yards between 2001 and 2016. His all-time high was 5,476 yards in 1 year. In his second highest year, he threw for 5,208 yards. How many passing yards did he throw in the remaining years?

Understand the problem. Then make a plan. How will you solve? Explain.

4. There are 86,400 seconds in 1 day. On most days, a student spends 28,800 seconds sleeping and 28,500 seconds in school. How many seconds are students awake, but *not* in school?

5. A pair of rhinoceroses weigh 14,860 pounds together. The female weighs 7,206 pounds. How much more does the male weigh than the female?

6. Earth is 24,873 miles around. If a person's blood vessels were laid out in a line, they would be able to circle Earth two times, plus 10,254 more miles. How many miles long are a person's blood vessels when laid out in a line?

7. Alaska has 22,041 more miles of shoreline than Florida and California combined. Alaska has 33,904 miles of shoreline. Florida has 8,436 miles of shoreline. How many miles of shoreline does California have?

Chapter 2 | Lesson 5

Think and Grow: Modeling Real Life

Example The attendance on the second day of a music festival is 10,013 fewer people than on the third day. How many total people attend the three-day music festival?

Day	Attendance
1	76,914
2	85,212
3	?

Think: What do you know? What do you need to find? How will you solve?

Step 1: How many people attend the festival on the third day?

```
| 85,212 | 10,013 |
|--------- a ---------|
```

a is the unknown sum.

$$85{,}212 + 10{,}013 = a$$

85,212
+ 10,013
_____

$a = $ _____

Remember, you can estimate 85,000 + 10,000 = 95,000 to check whether your answer is reasonable.

Step 2: Use a to find the total attendance for the three-day festival.

$$76{,}914 + 85{,}212 + a = f$$

f is the unknown sum.

$$76{,}914 + 85{,}212 + \underline{} = f$$

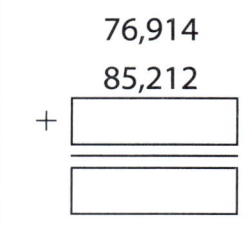

76,914
85,212
+ _____
_____

$f = $ _____

You can estimate 80,000 + 90,000 + 100,000 = 270,000 to check whether your answer is reasonable.

_____ total people attend the music festival.

Show and Grow   I can think deeper!

8. A construction company uses 3,239 more bricks to construct Building 1 than Building 2. How many bricks does the company use to construct all three buildings?

Building	Bricks Used
1	11,415
2	?
3	16,352

Name _____

Homework & Practice 2.5

Learning Target: Use the problem-solving plan to solve two-step addition and subtraction word problems.

Example A car owner wants to sell his car when the odometer reads 100,000 miles. The owner bought the car with 44,901 miles on it. He drives a total of 27,298 miles over the next two years. How many more miles can he drive before he sells his car?

Think: What do you know? What do you need to find? How will you solve?

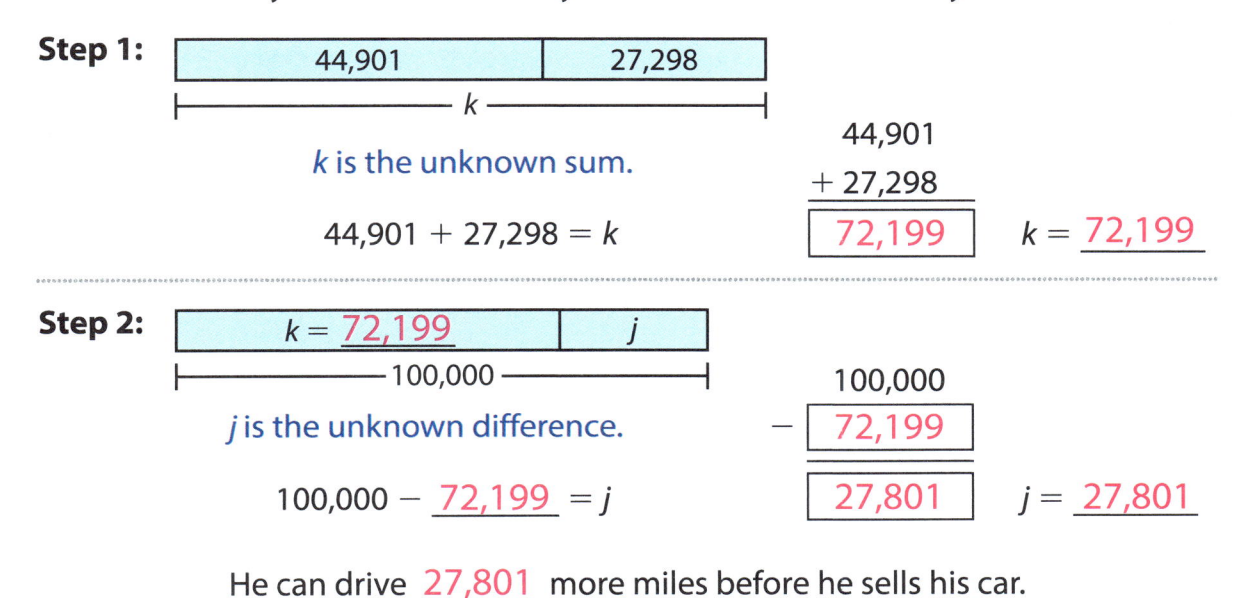

Step 1:

| 44,901 | 27,298 |

⊢——— k ———⊣

k is the unknown sum.

44,901 + 27,298 = k

$$\begin{array}{r} 44{,}901 \\ +\ 27{,}298 \\ \hline 72{,}199 \end{array}$$

k = __72,199__

Step 2:

| k = 72,199 | j |

⊢——— 100,000 ———⊣

j is the unknown difference.

100,000 − __72,199__ = j

$$\begin{array}{r} 100{,}000 \\ -\ 72{,}199 \\ \hline 27{,}801 \end{array}$$

j = __27,801__

He can drive __27,801__ more miles before he sells his car.

Understand the problem. Then make a plan. How will you solve? Explain.

1. A cargo plane weighs 400,000 pounds. After a load of cargo is removed, the plane weighs 336,985 pounds. Then a 12,395-pound load is removed. How many pounds of cargo are removed in all?

2. A ski resort uses 5,200 gallons of water per minute to make snow. A family uses 361 gallons of water each day. How many more gallons of water does the ski resort use to make snow in 2 minutes than a family uses in 1 day?

Chapter 2 | Lesson 5

3. In July, a website receives 379,162 fewer orders than in May and June combined. The website receives 542,369 orders in May and 453,708 orders in June. How many orders does the website receive in July?

4. **Writing** Write and solve a two-step word problem that can be solved using addition or subtraction.

5. **Modeling Real Life** World War I lasted from 1914 to 1918. World War II lasted from 1939 to 1945. How much longer did World War II last than World War I?

6. **Modeling Real Life** Twenty people each donate $9 to a charity. Sixty people each donate $8. The charity organizer wants to raise a total of $1,500. How much more money does the organizer need to raise?

7. **DIG DEEPER!** The blackpoll warbler migrates 2,376 miles, stops, and then flies another 3,289 miles to reach its destination. The arctic tern migrates 11,013 miles, stops, and then flies another 10,997 miles to reach its destination. How much farther is the arctic tern's migration than the blackpoll warbler's migration?

Blackpoll Warbler

Artic Tern

Review & Refresh

Write the time. Write another way to say the time.

8.

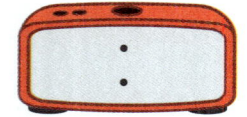

9.

10.

Name _____

Performance Task 2

1. The time line shows the population of Austin, Texas from 1995 to 2015.

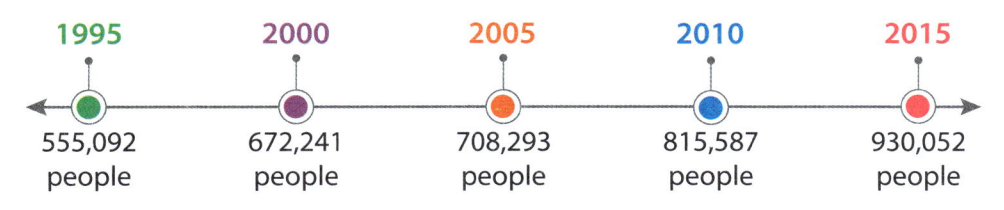

1995	2000	2005	2010	2015
555,092 people	672,241 people	708,293 people	815,587 people	930,052 people

_____ _____ _____ _____ _____

a. Which place would you round to when estimating population? Explain.

b. Estimate Austin's population each year on the timeline.

c. Use your estimates to complete the bar graph.

d. Between the years 1995 and 2015, did Austin's population increase, decrease, or stay the same? Explain.

e. During which five-year period did the population increase the most? Explain.

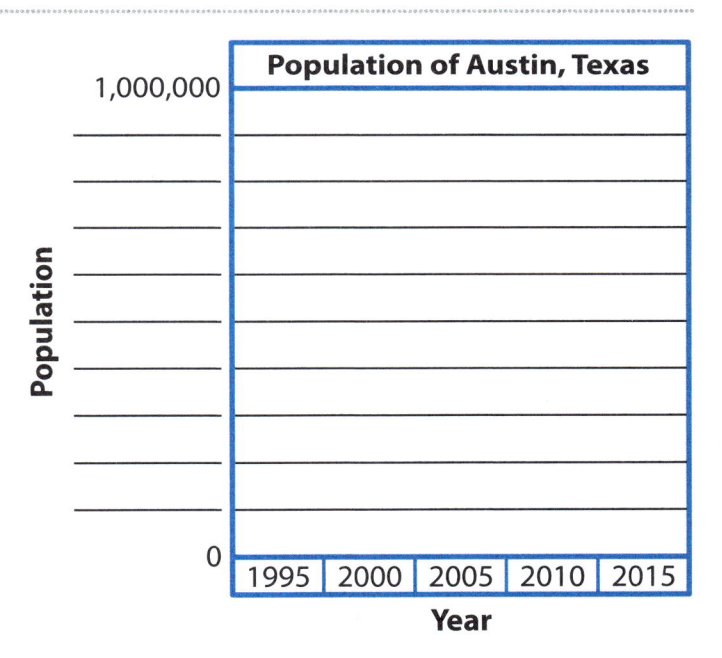

Population of Austin, Texas
(bar graph: Population vs Year — 1995, 2000, 2005, 2010, 2015; y-axis up to 1,000,000)

f. About how many more people lived in Austin in the year 2015 than in the year 1995?

g. Do you think the population will be more than 1,000,000 in the year 2020? Explain.

Chapter 2

63

Race to the Moon

Directions:
1. Players take turns.
2. On your turn, flip a Race for the Moon Card and find the sum or difference.
3. Move your piece to the next number on the board that is highlighted in your answer.

Name _____

Chapter Practice 2

2.1 Estimate Sums and Differences

Estimate the sum or difference.

1.   9,472
   − 2,863

2.   36,720
   + 34,529

3.   83,691
   − 50,466

4. 742,086 − 486,629 = _____

5. 216,987 + 72,429 = _____

2.2 Add Multi-Digit Numbers

Find the sum. Check whether your answer is reasonable.

6. Estimate: _____

     43,824
   + 9,452

7. Estimate: _____

     52,091
   + 27,869

8. Estimate: _____

     396,705
   + 212,629

9. **MP Logic** Find the missing digits.

      1 2 , ☐ 6 6
   +   9 , 9 4 ☐
   ─────────────
      2 2 , 5 1 1

      2 6 5 , 0 ☐ 1
   + 1 ☐ 7 , 4 8 9
   ─────────────
      4 5 2 , 5 2 0

Chapter 2   65

2.3 Subtract Multi-Digit Numbers

Find the difference. Then check your answer.

10. 13,246
   − 8,489

11. 32,700
   − 30,411

12. 268,790
   − 193,215

13. 973,287 − 8,345 = _____

14. 762,179 − 21,280 = _____

2.4 Use Strategies to Add or Subtract

Find the sum or difference. Then check your answer.

15. 13,500
   − 6,295

16. 914,562
   − 89,034

17. 630,254
   + 281,273

2.5 Problem Solving: Addition and Subtraction

18. **Modeling Real Life** A library has 12,850 books. There are 1,932 poetry books, 5,047 nonfiction books, and 2,891 graphic novels. The rest are fiction. How many fiction books does the library have?

19. **Modeling Real Life** A technology teacher wants to buy a 3-D printer for $3,495. He makes 3 payments of $999. How much money does the teacher have left to pay for the 3-D printer?

3 Multiply by One-Digit Numbers

- Have you ever seen a fireworks display? What types of events have fireworks displays?

- At a fireworks display, you see the lights before you hear the sounds because light travels faster than sound. How can you use multiplication to find out how far away you are from the fireworks?

Chapter Learning Target:
Understand multiplying one-digit numbers.

Chapter Success Criteria:
- I can find the product of two numbers.
- I can use rounding to estimate a product.
- I can write multiplication problems.
- I can solve a problem using an equation.

Name _____

3 Vocabulary

Review Words
Commutative Property of Multiplication
factors
product

Organize It

Use the review words to complete the graphic organizer.

[_____]

Changing the order of [_____] does not change the [_____].

$4 \times 3 = 12$ $3 \times 4 = 12$

So, $4 \times 3 = 3 \times 4$.

Define It

Use your vocabulary cards to match.

1. Distributive Property

   39
   × 7
   ─────
   [63]   7 × 9
   +[210] 7 × 30
   ─────
   273

2. partial products

   $3 \times (5 + 2) = (3 \times 5) + (3 \times 2)$

   $3 \times (5 - 2) = (3 \times 5) - (3 \times 2)$

68

Chapter 3 Vocabulary Cards

Distributive Property

partial products

The products found by breaking apart a factor into ones, tens, hundreds, and so on, and multiplying each of these by the other factor $$\begin{array}{r}39\\\times\ 7\\\hline\end{array}$$ partial products → $\boxed{63}$ $\ 7\times 9$ $+\boxed{210}$ $\ 7\times 30$ $\overline{273}$	$3\times(5+2)=(3\times 5)+(3\times 2)$ $3\times(5-2)=(3\times 5)-(3\times 2)$

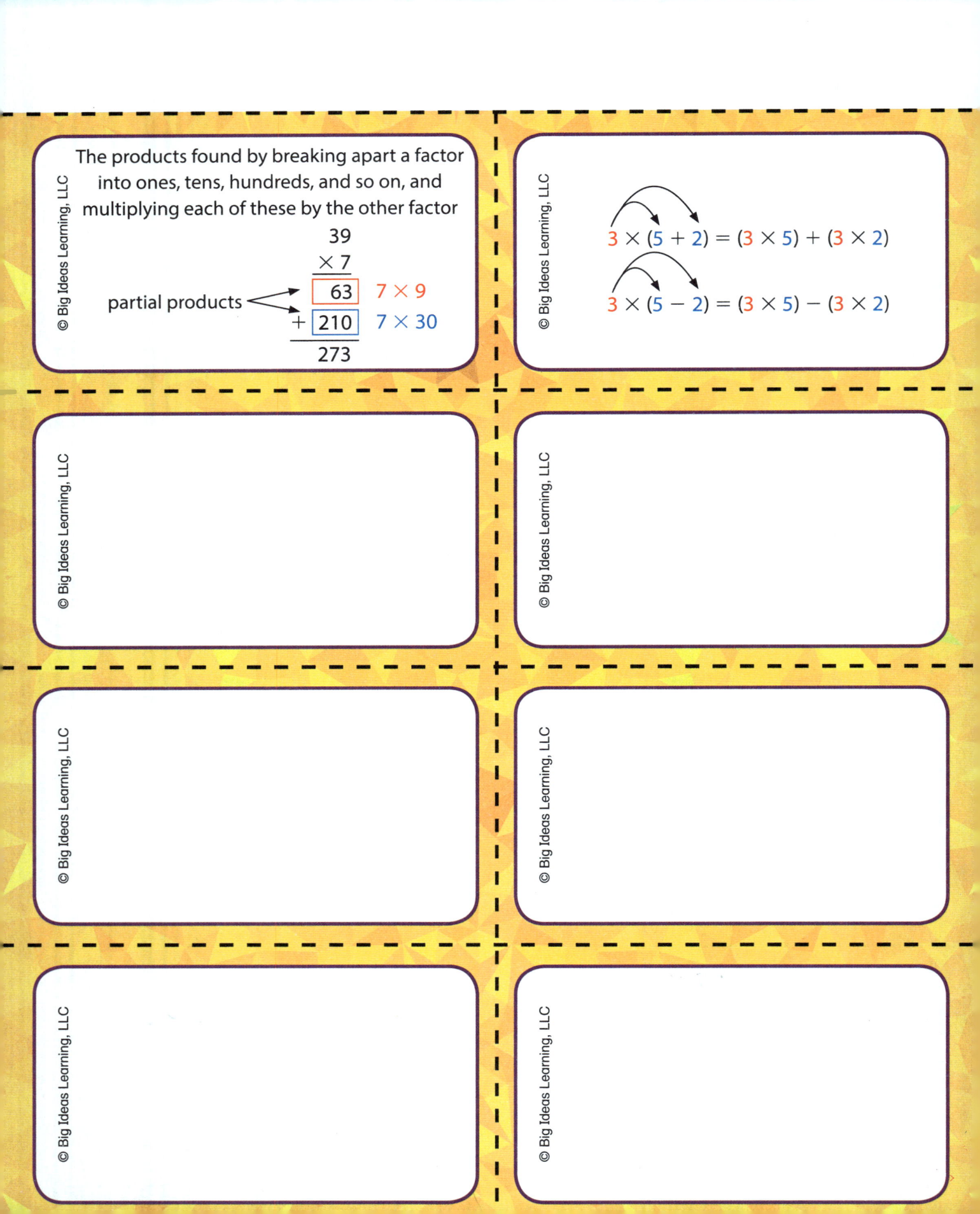

Name _____

Multiply Tens, Hundreds, and Thousands 3.2

Learning Target: Use place value to multiply by tens, hundreds, or thousands.

Success Criteria:
- I can find the product of a one-digit number and a multiple of ten, one hundred, or one thousand.
- I can describe a pattern when multiplying by tens, hundreds, or thousands.

Explore and Grow

Use models to find each product. Draw your models.

4 × 3 = _12_	4 × 30 = _120_
4 × 300 = _1,200_	4 × 3,000 = _12,000_

What pattern do you notice?

MP Repeated Reasoning How does 3 × 7 help you to find 3 × 7,000? Explain.

3 × 7 = 21   3 × 7,000 = 21,000

Chapter 3 | Lesson 2   75

Think and Grow: Multiply Tens, Hundreds, and Thousands

You can use place value to multiply by tens, hundreds, or thousands.

Example Find each product.

7 × 200 = 7 × _____ hundreds

= _____ hundreds

= _____

So, 7 × 200 = _____.

3 × 4,000 = 3 × _____ thousands

= _____ thousands

= _____

So, 3 × 4,000 = _____.

Example Find each product.

8 × 5 = 40   Multiplication fact

8 × 50 = 400   Find 8 × 5; write 1 zero to show tens.

8 × 500 = _____   Find 8 × 5; write 2 zeros to show hundreds.

8 × 5,000 = _____   Find 8 × 5; write 3 zeros to show thousands.

Notice the Pattern: Write the multiplication fact and the same number of zeros that are in the second factor.

Show and Grow   I can do it!

Find each product.

1. 6 × 9 = 54
   6 × 90 = 540
   6 × 900 = 5400
   6 × 9,000 = 54,000

2. 5 × 2 = 10
   5 × 20 = 100
   5 × 200 = 500
   5 × 2,000 = 10,000

3. 2 × 3 = 6
   2 × 300 = 600

4. 9 × 8 = 72
   9 × 8,000 = 72,000

Name _____

Estimate Products by Rounding 3.3

Learning Target: Use rounding to estimate products.
Success Criteria:
- I can use rounding to estimate a product.
- I can find two estimates that a product is between.
- I can tell whether a product is reasonable.

Explore and Grow

There are 92 marbles in each jar.

Estimate the total number of marbles in two ways.

Round 92 to the nearest ten.   Round 92 to the nearest hundred.

Which estimate do you think is closer to the total number of marbles? Explain.

MP **Repeated Reasoning** Would your answer to the question above change if there were 192 marbles in each jar? Explain.

Chapter 3 | Lesson 3   81

Think and Grow: Estimate Products

You can estimate a product by rounding.

Example Estimate 7 × 491.

Round 491 to the nearest hundred. Then multiply.

7 × 491
↓
7 × _____ = _____

So, 7 × 491 is about _____.

> Remember, an estimate is a number that is close to an exact number.

When solving multiplication problems, you can check whether an answer is reasonable by finding two estimates that a product is between.

Example Find two estimates that the product of 4 × 76 is between.

Think: 76 is between 70 and 80.

4 × 76   4 × 76
↓   ↓
4 × 70 = _____   4 × 80 = _____

So, the product is between _____ and _____.

Show and Grow   I can do it!

Estimate the product.

1. 3 × 89

2. 8 × 721

3. 5 × 7,938

Find two estimates that the product is between.

4. 9 × 44

5. 2 × 657

6. 6 × 4,243

Name _____

Learning Target: Use the Distributive Property to multiply.
Success Criteria:
- I can draw an area model to multiply.
- I can use known facts to find a product.
- I can explain how to use the Distributive Property.

Use the Distributive Property to Multiply

3.4

Explore and Grow

Use base ten blocks to model 4 × 16. Draw your model. Then find the area of the model.

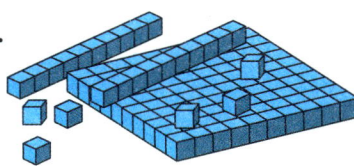

4 × 16 = _____

Break apart 16 to show two smaller models. Find the area of each model. What do you notice about the sum of the areas?

Area = _____   Area = _____

MP Reasoning How does this strategy relate to the Distributive Property? Explain.

Chapter 3 | Lesson 4   87

Think and Grow: Use the Distributive Property to Multiply

One way to multiply a two-digit number is to first break apart the number. Then use the Distributive Property.

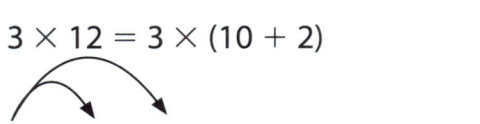

Think: 12 = 10 + 2

$3 \times 12 = 3 \times (10 + 2)$

$3 \times (10 + 2) = (3 \times 10) + (3 \times 2)$ **Distributive Property**

Example   Use an area model to find 6×18.

Model the expression. Break apart 18 as $10 + 8$.

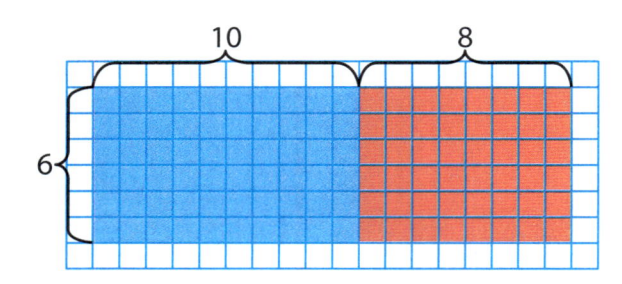

$6 \times 18 = 6 \times (\underline{} + \underline{})$ Rewrite 18 as $10 + 8$.

$ = (6 \times 10) + (6 \times 8)$ Distributive Property

$ = 60 + 48$

$ = \underline{}$

Show and Grow   I can do it!

Draw an area model. Then find the product.

1. $7 \times 11 = \underline{}$

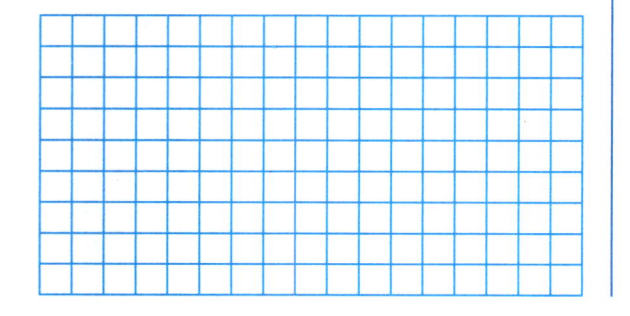

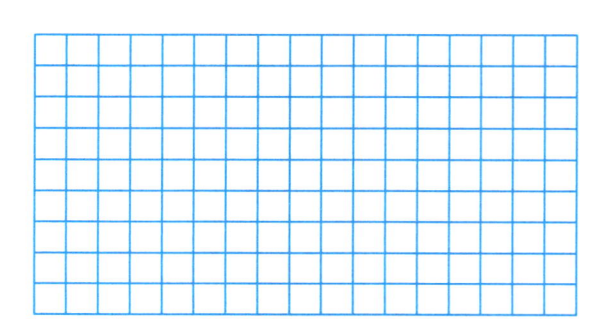

2. $2 \times 15 = \underline{}$

88

Name _____

Apply and Grow: Practice

Draw an area model. Then find the product.

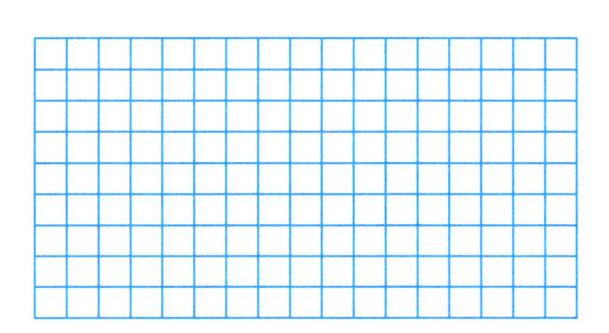

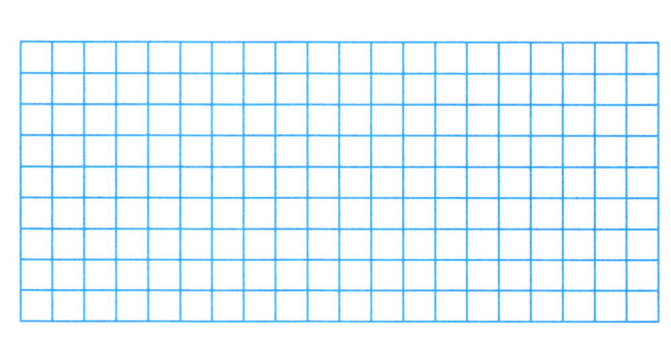

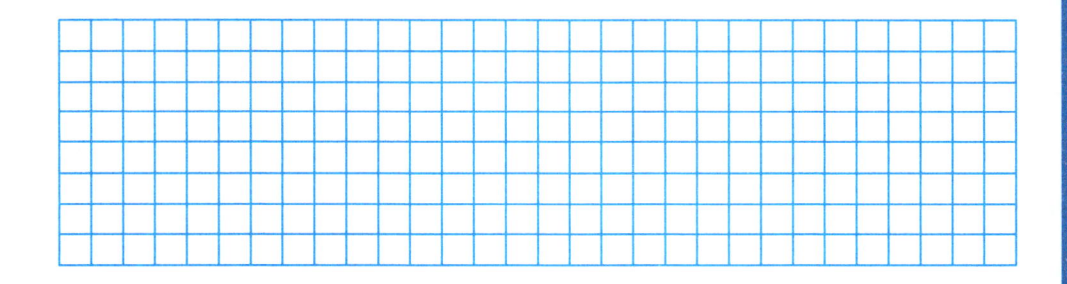

3. 4 × 16 = _____

4. 9 × 18 = _____

5. 6 × 27 = _____

Find the product.

6. 3 × 46 = _____

7. 8 × 35 = _____

8. 5 × 72 = _____

9. **Number Sense** Use the area model to complete the equation.

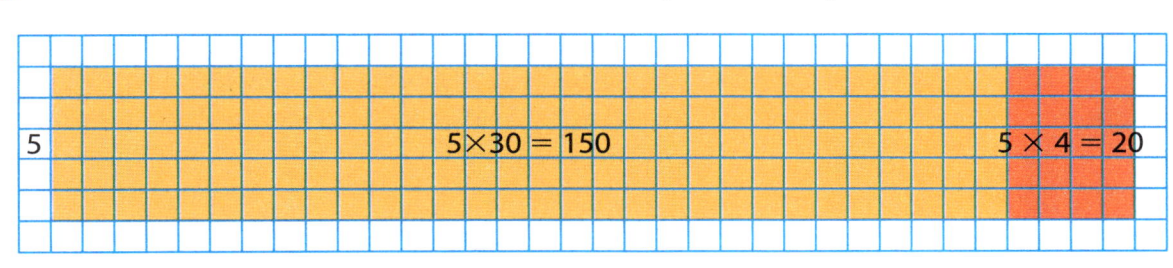

5 | 5 × 30 = 150 | 5 × 4 = 20

5 × _____ = _____

Chapter 3 | Lesson 4

89

Think and Grow: Modeling Real Life

Example A parking garage has 9 floors. Each floor has 78 parking spaces. 705 cars are trying to park in the garage. Are there enough parking spaces? Explain.

Multiply the number of floors by the number of parking spaces on each floor.

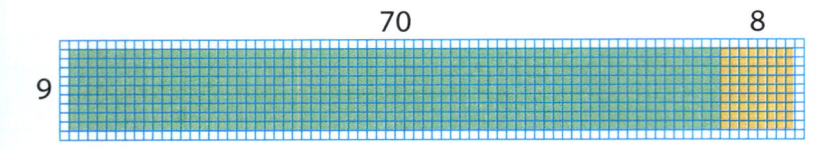

$9 \times 78 = 9 \times (70 + 8)$

$= (9 \times 70) + (9 \times 8)$

$= \underline{} + \underline{}$

$= \underline{}$

Compare the number of spaces to the number of cars trying to park in the garage.

There _____ enough parking spaces.

Explain:

Show and Grow — I can think deeper!

10. Your piano teacher wants you to practice playing the piano 160 minutes this month. So far, you have practiced 5 minutes each day for 24 days. Have you reached the goal your teacher set for you? Explain.

11. Newton has $150. He wants to buy a bicycle that costs 4 times as much as the helmet. Does he have enough money to buy the bicycle and the helmet? Explain.

$28

12. A baby orangutan weighs 3 pounds. The mother orangutan weighs 27 times as much as the baby. The father orangutan weighs 63 times as much as the baby. What are the weights of the mother and the father?

Name _____

Homework & Practice 3.4

Learning Target: Use the Distributive Property to multiply.

Example Use an area model to find 8 × 24.

Model the expression. Break apart 24 as 20 + 4.

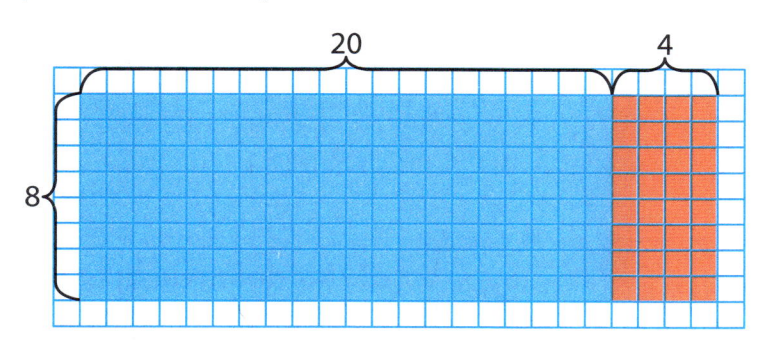

8 × 24 = 8 × (__20__ + __4__)   Rewrite 24 as 20 + 4.

= (8 × 20) + (8 × 4)   Distributive Property

= 160 + 32

= __192__

Draw an area model. Then find the product.

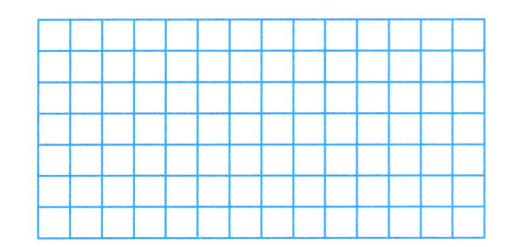

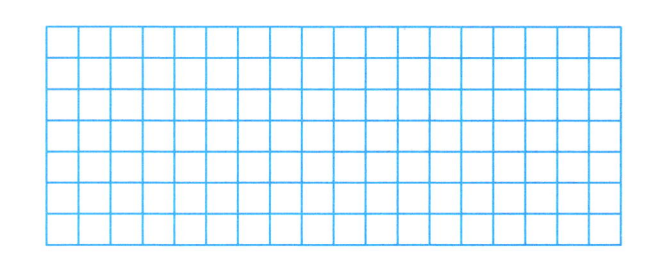

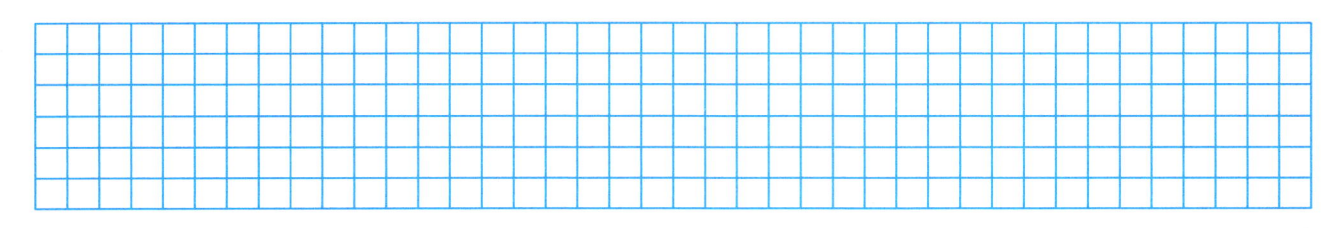

1. 3 × 12 = _____

2. 5 × 16 = _____

3. 4 × 34 = _____

Chapter 3 | Lesson 4

Find the product.

4. 9 × 56 = _____

5. 71 × 2 = _____

6. 3 × 77 = _____

7. Writing Explain how you can use the Distributive Property to find a product.

8. Structure Use the Distributive Property to find 6 × 18 two different ways.

9. Reasoning To find 4 × 22, would you rather break apart the factor 22 as 20 + 2 or as 11 + 11? Explain.

10. Modeling Real Life The diameter of a firework burst, in feet, is 45 times the height of the shell in inches. You have an 8-inch firework shell. Will the shell produce a firework burst that has a diameter greater than 375 feet? Explain.

11. Modeling Real Life A juvenile bearded dragon should eat 48 crickets each day. You have 150 crickets. Do you have enough crickets to feed 3 juvenile bearded dragons? Explain.

Review & Refresh

Plot the fraction on a number line.

12. $\frac{7}{4}$

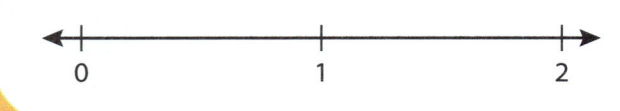

13. $\frac{4}{3}$

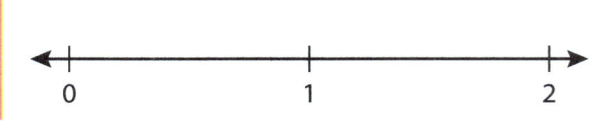

Name _____

Use Expanded Form to Multiply 3.5

Learning Target: Use expanded form and the Distributive Property to multiply.

Success Criteria:
- I can use an area model to multiply.
- I can use expanded form and the Distributive Property to find a product.

Explore and Grow

Write 128 in expanded form.

128 = _____ + _____ + _____

Use expanded form to label the area model for 5 × 128. Find the area of each part.

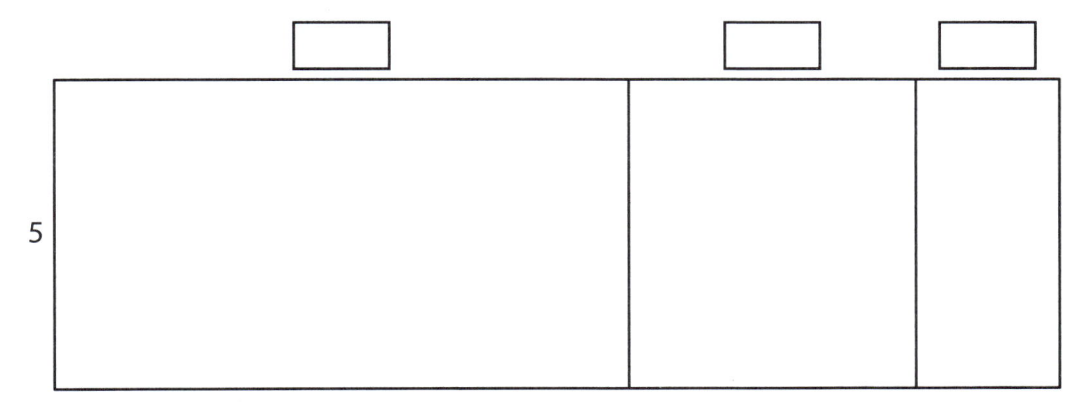

[Area model with height 5, divided into three parts with small boxes above each part for labels]

What is the sum of all of the parts? How does the sum relate to the product of 5 and 128?

MP Repeated Reasoning Explain to your partner how you can use expanded form to find 364 × 8.

Chapter 3 | Lesson 5

Think and Grow: Use Expanded Form to Multiply

You can use expanded form and the Distributive Property to multiply.

Example Find 8×74.

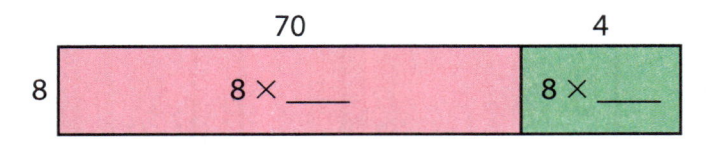

	70	4
8	8 × ___	8 × ___

$8 \times 74 = 8 \times (70 + 4)$ Write 74 in expanded form.

$ = (8 \times 70) + (8 \times 4)$ Distributive Property

$ = \underline{} + \underline{}$

$ = \underline{}$ So, $8 \times 74 = \underline{}$.

Example Find $2 \times 5{,}607$.

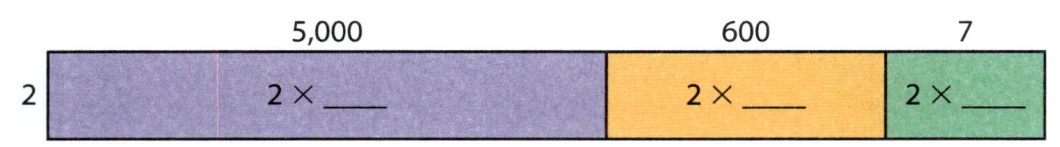

	5,000	600	7
2	2 × ___	2 × ___	2 × ___

$2 \times 5{,}607 = 2 \times (5{,}000 + 600 + 7)$ Write 5,607 in expanded form.

$\phantom{2 \times 5{,}607} = (2 \times 5{,}000) + (2 \times 600) + (2 \times 7)$ Distributive Property

$\phantom{2 \times 5{,}607} = \underline{} + \underline{} + \underline{}$

$\phantom{2 \times 5{,}607} = \underline{}$ So, $2 \times 5{,}607 = \underline{}$.

Show and Grow   I can do it!

Find the product.

1. $4 \times 306 = 4 \times (\underline{} + \underline{})$

    $ = (4 \times \underline{}) + (4 \times \underline{})$

    $ = \underline{} + \underline{}$

    $ = \underline{}$

2. 7×549

Name _____

Apply and Grow: Practice

Find the product.

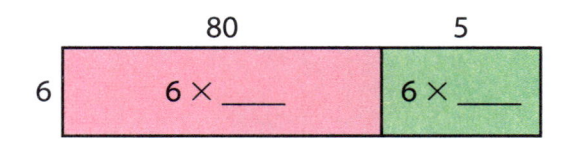

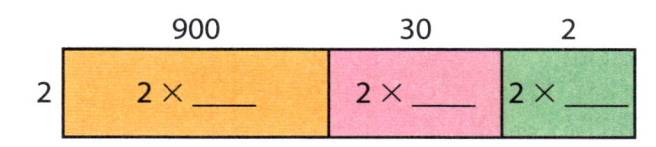

3. 6 × 85 = _____

	80	5
6	6 × ____	6 × ____

4. 2 × 932 = _____

	900	30	2
2	2 × ____	2 × ____	2 × ____

5. 4 × 690 = 4 × (_____ + _____)

= (4 × _____) + (4 × _____)

= _____ + _____

= _____

6. 1,027 × 9 = (_____ + _____ + _____) × 9

= (_____ × 9) + (_____ × 9) + (_____ × 9)

= _____ + _____ + _____

= _____

7. 487 × 5 = _____

8. 8 × 2,483 = _____

9. A basketball player made 269 three-point shots in a season. How many points did he score from three-point shots?

10. Your cousin runs 6 miles each week. There are 5,280 feet in a mile. How many feet does your cousin run each week?

11. **YOU BE THE TEACHER** Your friend finds 744 × 3. Is your friend correct? Explain.

744 × 3 = (700 + 40 + 4) × 3

= (700 × 3) + (40 × 3) + (4 × 3)

= 2,100 + 120 + 12

= 2,232

12. **DIG DEEPER!** What is the greatest possible product of a two-digit number and a one-digit number? Explain.

Chapter 3 | Lesson 5

Think and Grow: Modeling Real Life

Example A baby hippo is fed 77 fluid ounces of milk 5 times each day. There are 128 fluid ounces in 1 gallon. Are 2 gallons of milk enough to feed the baby hippo for 1 day?

Find the number of fluid ounces of milk the baby hippo drinks in 1 day.

$5 \times 77 = 5 \times (70 + 7)$

$= (5 \times 70) + (5 \times 7)$

$= \underline{} + \underline{}$

$= \underline{}$

Find the number of fluid ounces in 2 gallons.

$2 \times 128 = 2 \times (100 + 20 + 8)$

$= (2 \times 100) + (2 \times 20) + (2 \times 8)$

$= \underline{} + \underline{} + \underline{}$

$= \underline{}$

Compare the products.

So, 2 gallons of milk _____ enough to feed the baby hippo for 1 day.

Show and Grow I can think deeper!

13. A school with 6 grades goes on a field trip. There are 64 students in each grade. One bus holds 48 students. Will 8 buses hold all of the students?

14. The average life span of a firefly is 61 days. The average life span of a Monarch butterfly is 4 times as long as that of a firefly. How many days longer is the average lifespan of a Monarch butterfly than a firefly?

15. A tourist is in Denver. His car can travel 337 miles using 1 tank of gasoline. He wants to travel to a city using no more than 3 tanks of gasoline. To which cities could the tourist travel?

City	Distance from Denver
Los Angeles	1,016 miles
Phoenix	821 miles
Dallas	796 miles

Name _____

Homework & Practice 3.5

Learning Target: Use expanded form and the Distributive Property to multiply.

Example Find 5×391.

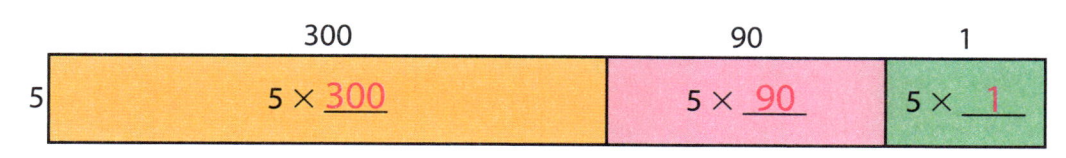

	300	90	1
5	$5 \times \underline{300}$	$5 \times \underline{90}$	$5 \times \underline{1}$

$5 \times 391 = 5 \times (300 + 90 + 1)$ Write 391 in expanded form.

$\qquad = (5 \times 300) + (5 \times 90) + (5 \times 1)$ Distributive Property

$\qquad = \underline{1{,}500} + \underline{450} + \underline{5}$

$\qquad = \underline{1{,}955}$

So, $5 \times 391 = \underline{1{,}955}$.

Find the product.

1. $7 \times 803 = $ _____

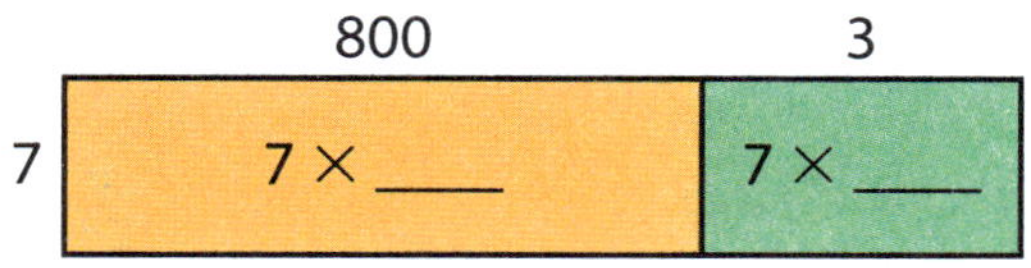

	800	3
7	$7 \times $ ___	$7 \times $ ___

2. $9 \times 1{,}024 = $ _____

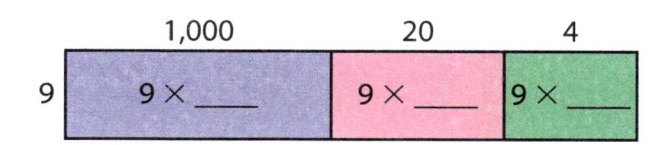

	1,000	20	4
9	$9 \times $ ___	$9 \times $ ___	$9 \times $ ___

3. $43 \times 8 = (40 + 3) \times 8$

$\qquad = (\underline{} \times 8) + (\underline{} \times 8)$

$\qquad = \underline{} + \underline{}$

$\qquad = \underline{}$

4. $4 \times 742 = 4 \times (\underline{} + \underline{} + \underline{})$

$\qquad = (4 \times \underline{}) + (4 \times \underline{}) + (4 \times \underline{})$

$\qquad = \underline{} + \underline{} + \underline{}$

$\qquad = \underline{}$

5. $3 \times 482 = $ _____

6. $4{,}591 \times 6 = $ _____

Chapter 3 | Lesson 5

7. Each lobe on a Venus flytrap has 6 trigger hairs that sense and capture insects. A Venus flytrap has 128 lobes. How many trigger hairs does it have?

8. A human skeleton has 206 bones. How many bones do 4 skeletons have in all?

9. Writing Explain how you can find $5 \times 7{,}303$ using expanded form.

10. Structure Rewrite the expression as a product of two factors.

$$(6{,}000 \times 3) + (70 \times 3) + (4 \times 3)$$

11. Modeling Real Life A summer camp has 8 different groups of students. There are 35 students in each group. There are 24 shirts in a box. Will 9 boxes be enough for each student to get one shirt?

12. Modeling Real Life Firefighters respond to 65 calls in 1 week. Police officers respond to 8 times as many calls as firefighters in the same week. How many more calls do police officers respond to than firefighters in that week?

Review & Refresh

Write the total mass shown.

13. 500 g, 100 g, 100 g, 100 g, 100 g, 1 g, 1 g, 1 g, 10 g

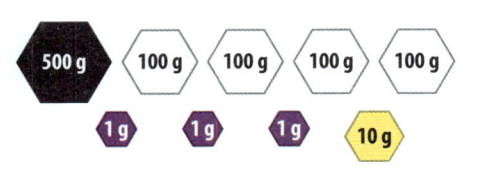

14. 1 kg, 5 g, 100 g, 100 g, 100 g, 10 g, 10 g, 10 g, 10 g, 1 g

98

Name _____

Use Partial Products to Multiply 3.6

Learning Target: Use place value and partial products to multiply.

Success Criteria:
- I can use place value to tell the value of each digit in a number.
- I can write the partial products for a multiplication problem.
- I can add the partial products to find a product.

Explore and Grow

Use the area model to find 263 × 4.

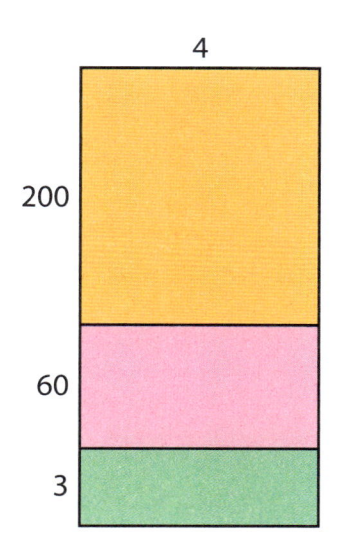

```
      4
  ┌───────┐
  │       │
200│       │
  │       │
  ├───────┤        263
 60│       │       ×  4
  ├───────┤       ┌─────┐
  3│       │      │     │
  └───────┘       ├─────┤
                  │     │
                 +├─────┤
                  │     │
                  └─────┘
```

MP Repeated Reasoning Does the product change if you multiply the ones first, then the tens and hundreds? Explain.

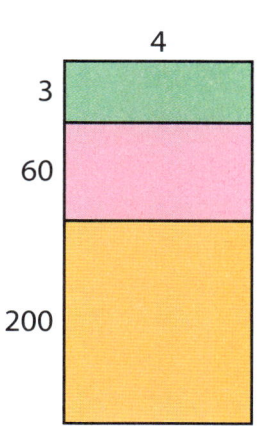

```
      4
  ┌───────┐
  3│       │
  ├───────┤
 60│       │
  ├───────┤
  │       │
200│       │
  │       │
  └───────┘
```

Chapter 3 | Lesson 6   99

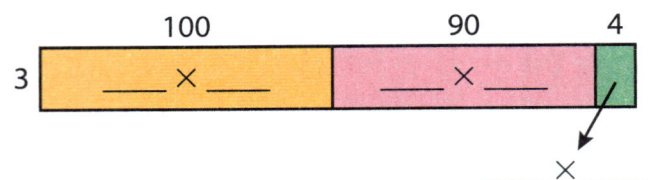

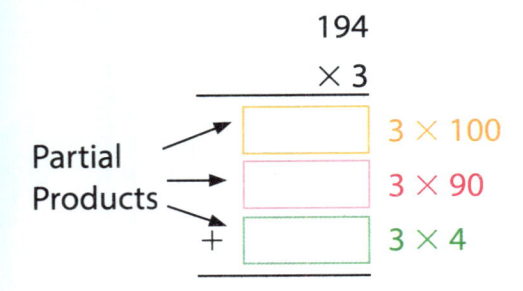

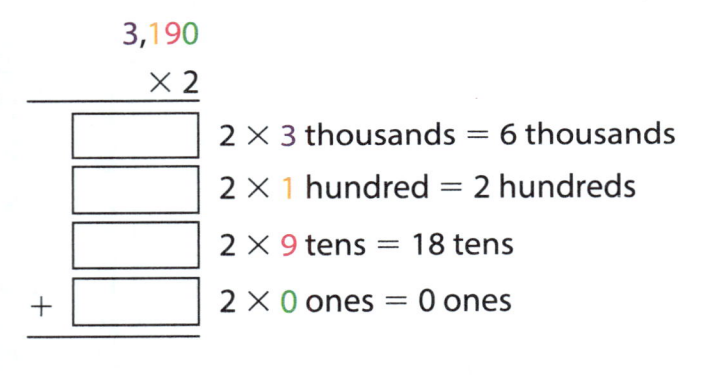

Think and Grow: Use Partial Products to Multiply

Partial products are found by breaking apart a factor into ones, tens, hundreds, and so on, and multiplying each of these by the other factor.

Example Use an area model and partial products to find 194 × 3.

```
    194
  ×   3
```

Partial Products
- 3 × 100
- 3 × 90
- + 3 × 4

_____ Add the partial products.

| 100 | 90 | 4 |
3 | __ × __ | __ × __ | __ × __ |

__ × __

So, 194 × 3 = _____.

Example Use place value and partial products to find 3,190 × 2.

```
   3,190
  ×    2
```

- 2 × 3 thousands = 6 thousands
- 2 × 1 hundred = 2 hundreds
- 2 × 9 tens = 18 tens
- + 2 × 0 ones = 0 ones

_____ Add the partial products.

So, 3,190 × 2 = _____.

You can find partial products in any order.

Show and Grow I can do it!

Find the product.

1.
```
      86
  ×    5
```

2.
```
     502
  ×    7
```

3.
```
   5,367
  ×    4
```

100

Name _____

Homework & Practice 3.6

Learning Target: Use place value and partial products to multiply.

Example Use place value and partial products to find 4,935 × 4.

```
    4,935
  ×     4
  ──────
   16,000    4 × 4 thousands = 16 thousands
    3,600    4 × 9 hundreds = 36 hundreds
      120    4 × 3 tens = 12 tens
 +     20    4 × 5 ones = 20 ones
  ──────
   19,740   Add the partial products.
```

So, 4,935 × 4 = __19,740__.

Find the product.

1. 37
    ×   5

2. 781
    ×   9

3. 2,937
    ×   3

4. 782
    ×   6

5. 85
    × 8

6. 8,026
    ×   2

Chapter 3 | Lesson 6 103

Find the product.

7. 399
    × 7

8. 7,390
    × 4

9. 9,286
    × 5

10. **Number Sense** Which four numbers are the partial products that you add to find the product of 3,472 and 6?

    18,000   2,400   1,800

    420   240   12

11. **DIG DEEPER!** The sum of a four-digit number and a one-digit number is 7,231. The product of the numbers is 28,908. What are the numbers?

12. **Modeling Real Life** The height of the Eiffel Tower is 38 feet more than 3 times the height of Big Ben. What is the height of the Eiffel Tower?

    316 ft

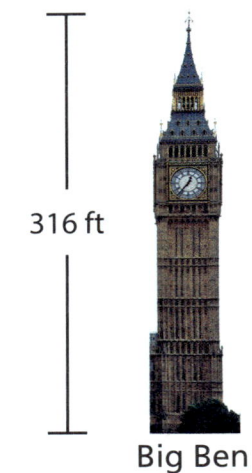

    Big Ben

13. **Modeling Real Life** An animal shelter owner has 9 dogs and 4 cats ready for adoption. How much money will the owner collect when all of the animals are adopted?

    Adopt a Pet Today!
    Cats: $450
    Dogs: $1,150

Review & Refresh

Write all of the names for the quadrilateral.

14.

15.

104

Name _____

Learning Target: Multiply two-digit numbers by one-digit numbers.
Success Criteria:
- I can multiply to find the partial products.
- I can show 10 ones regrouped as 1 ten.
- I can find the product.

Multiply Two-Digit Numbers by One-Digit Numbers 3.7

Explore and Grow

Model 24 × 3. Draw to show your model. Then find the product.

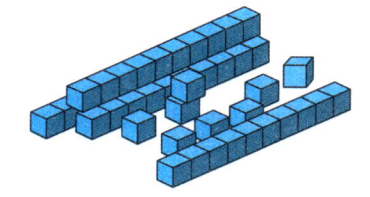

24 × 3 = _____

How did you use regrouping to find the product?

MP Precision Use a model to find 15 × 8.

Chapter 3 | Lesson 7

105

Think and Grow: Use Regrouping to Multiply

Example Find 32 × 6.

Estimate: 30 × 6 = _____

Step 1: Multiply the ones. Regroup.

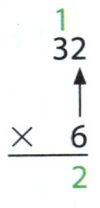

```
  1
  32
×  6
-----
   2
```

- 6 × 2 ones = 12 ones
- Regroup 12 ones as 1 ten and 2 ones.

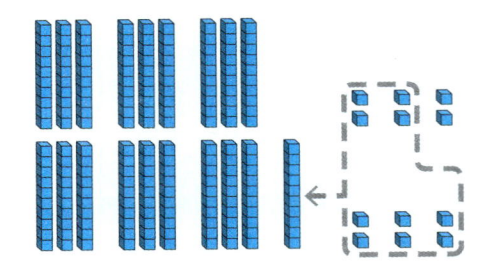

Step 2: Multiply the tens. Add any regrouped tens.

```
  1
  32
×  6
-----
 192
```

- 6 × 3 tens = 18 tens
- 18 tens + 1 ten = 19 tens, or 1 hundred and 9 tens

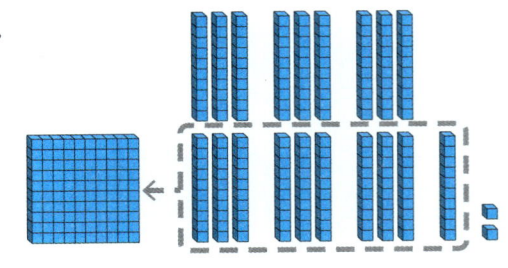

So, 32 × 6 = _____.

Check: Because _____ is close to the estimate, _____, the answer is reasonable.

Show and Grow   I can do it!

1. Use the model to find the product.

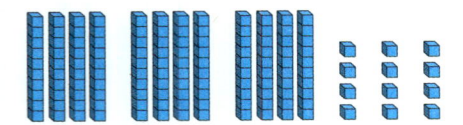

   3 × 44 = _____

Find the product. Check whether your answer is reasonable.

2. Estimate: _____

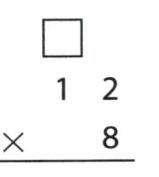

```
   1 2
 ×   8
------
```

3. Estimate: _____

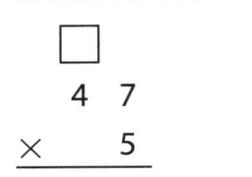

```
   4 7
 ×   5
------
```

4. Estimate: _____

```
   2 3
 ×   6
------
```

106

Name _____

Homework & Practice 3.7

Learning Target: Multiply two-digit numbers by one-digit numbers.

Example Find 64 × 5.

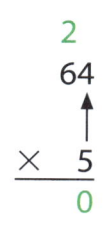

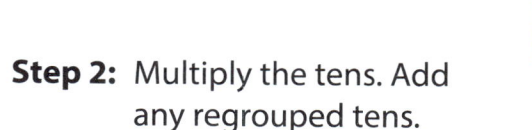

Estimate: 60 × 5 = **300**

Step 1: Multiply the ones. Regroup.

```
  2
 64
× 5
———
  0
```

Step 2: Multiply the tens. Add any regrouped tens.

```
  2
 64
× 5
———
320
```

So, 64 × 5 = **320**.

Check: Because **320** is close to the estimate, **300**, the answer is reasonable.

1. Use the model to find the product.

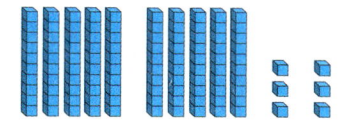

2 × 53 = _____

Find the product. Check whether your answer is reasonable.

2. Estimate: _____

```
  2 4
×   3
—————
```

3. Estimate: _____

```
  1 8
×   7
—————
```

4. Estimate: _____

```
  3 3
×   6
—————
```

5. Estimate: _____

```
 47
× 9
————
```

6. Estimate: _____

```
 65
× 4
————
```

7. Estimate: _____

```
 72
× 8
————
```

Chapter 3 | Lesson 7

109

Find the product. Check whether your answer is reasonable.

8. Estimate: _____

49 × 5 = _____

9. Estimate: _____

7 × 86 = _____

10. Estimate: _____

93 × 3 = _____

11. You read 56 pages each week. How many pages do you read in 8 weeks?

12. **Number Sense** The sum of two numbers is 20. The product of the two numbers is 51. What are the two numbers?

13. **Reasoning** Your friend multiplies 58 by 6 and says that the product is 3,048. Is your friend's answer reasonable? Explain.

14. **DIG DEEPER!** How much greater is 4 × 26 than 3 × 26? Explain how you know without multiplying.

15. **Modeling Real Life** A self-balancing scooter travels 12 miles per hour. An all-terrain vehicle can travel 6 times as fast as the scooter. A go-kart can travel 67 miles per hour. Which vehicle can travel the fastest?

Self-balancing scooter

16. **Modeling Real Life** It takes a spaceship 3 days to reach the moon from Earth. It takes a spaceship 14 times as many days to reach Mars from Earth. How long would it take the spaceship to travel from Earth to Mars and back?

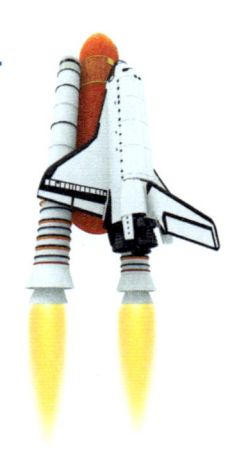

Review & Refresh

17. You spend 21 fewer minutes riding the bus to school than getting ready in the morning. You take 36 minutes to get ready. How long is the bus ride?

Name _____

Multiply Three- and Four-Digit Numbers by One-Digit Numbers

3.8

Learning Target: Multiply multi-digit numbers by one-digit numbers.

Success Criteria:
- I can multiply to find the partial products.
- I can show how to regroup more than 10 tens.
- I can find the product.

Explore and Grow

Use any strategy to find each product.

$7 \times 39 =$ _____

$7 \times 439 =$ _____

MP Structure How are the equations the same? How are they different?

Chapter 3 | Lesson 8

111

Think and Grow: Use Regrouping to Multiply

Example Find 795 × 4.

Remember, you can estimate 800 × 4 = 3,200 to check whether the answer is reasonable.

Step 1: Multiply the ones. Regroup.

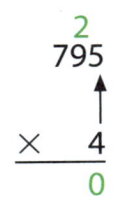

```
   2
  795
×   4
─────
    0
```

- 4 × 5 ones = 20 ones
- Regroup 20 ones as 2 tens and 0 ones.

Step 2: Multiply the tens. Add any regrouped tens.

```
  32
  795
×   4
─────
   80
```

- 4 × 9 tens = 36 tens
- 36 tens + 2 tens = 38 tens; Regroup 38 tens as 3 hundreds and 8 tens.

Step 3: Multiply the hundreds. Add any regrouped hundreds.

```
  32
  795
×   4
─────
 3,180
```

- 4 × 7 hundreds = 28 hundreds
- 28 hundreds + 3 hundreds = 31 hundreds, or 3 thousands and 1 hundred

So, 795 × 4 = _____.

Example Find 6,084 × 2.

Estimate: 6,000 × 2 = _____

```
  6,084
×     2
```

Multiply the ones, tens, hundreds, and thousands by 2. Regroup as necessary.

Check: Because _____ is close to the estimate, _____, the answer is reasonable.

Show and Grow   I can do it!

Find the product. Check whether your answer is reasonable.

1. Estimate: _____

```
  123
×   5
```

2. Estimate: _____

```
  907
×   3
```

3. Estimate: _____

```
  7,315
×     6
```

112

Name _____

Homework & Practice 3.8

Learning Target: Multiply multi-digit numbers by one-digit numbers.

Example Find 4,932 × 6.

Estimate: 5,000 × 6 = __30,000__

Step 1: Multiply the ones. Regroup.

$$\begin{array}{r} \overset{1}{4,932} \\ \times 6 \\ \hline 2 \end{array}$$

Step 2: Multiply the tens. Add any regrouped tens.

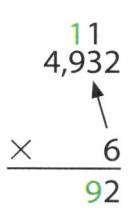

$$\begin{array}{r} \overset{11}{4,932} \\ \times 6 \\ \hline 92 \end{array}$$

Step 3: Multiply the hundreds. Add any regrouped hundreds.

$$\begin{array}{r} \overset{5\,11}{4,932} \\ \times 6 \\ \hline 592 \end{array}$$

Step 4: Multiply the thousands. Add any regrouped thousands.

$$\begin{array}{r} \overset{5\,11}{4,932} \\ \times 6 \\ \hline 29,592 \end{array}$$

So, 4,932 × 6 = __29,592__.

Check: Because __29,592__ is close to the estimate, __30,000__, the answer is reasonable.

Find the product. Check whether your answer is reasonable.

1. Estimate: _____

$$\begin{array}{r} 304 \\ \times 9 \\ \hline \end{array}$$

2. Estimate: _____

$$\begin{array}{r} 617 \\ \times 8 \\ \hline \end{array}$$

3. Estimate: _____

$$\begin{array}{r} 5,939 \\ \times 4 \\ \hline \end{array}$$

4. Estimate: _____

$$\begin{array}{r} 2,754 \\ \times 3 \\ \hline \end{array}$$

5. Estimate: _____

$$\begin{array}{r} 8,465 \\ \times 6 \\ \hline \end{array}$$

6. Estimate: _____

$$\begin{array}{r} 822 \\ \times 7 \\ \hline \end{array}$$

Chapter 3 | Lesson 8

Find the product. Check whether your answer is reasonable.

7. Estimate: _____

629 × 5 = _____

8. Estimate: _____

7 × 1,836 = _____

9. Estimate: _____

453 × 3 = _____

Compare.

10. 3 × 6,782 ◯ 2 × 3,391

11. 1,392 × 9 ◯ 2,493 × 4

12. A backpacker hikes the Buckeye Trail 5 times. The trail is 1,444 miles long. How many miles has he hiked on the Buckeye Trail in all?

13. **Number Sense** What number is 980 more than the product of 6,029 and 8?

14. **YOU BE THE TEACHER** Newton says that the product of a three-digit number and a one-digit number is always a three-digit number. Is Newton correct? Explain.

15. **Modeling Real Life** The numbers of songs Newton and Descartes download are shown. You download 2 times as many songs as Descartes. Who downloads the most songs?

Newton: 351 songs

Descartes: 167 songs

Review & Refresh

Find the sum or difference. Use the inverse operation to check.

16. 847
     − 162

17. 612
     + 289

18. 500
     − 351

116

Name _____

Use Properties to Multiply 3.9

Learning Target: Use properties to multiply.
Success Criteria:
- I can use the Commutative Property of Multiplication to multiply.
- I can use the Associative Property of Multiplication to multiply.
- I can use the Distributive Property to multiply.

Explore and Grow

Use any strategy to find each product. Explain the strategy you used to find each product.

$7 \times 25 \times 4 =$ _____	$5 \times 79 \times 2 =$ _____
$9 \times 5,001 =$ _____	$9 \times 4,999 =$ _____

MP **Construct Arguments** Compare your strategies with your partner's. How are they alike? How are they different?

Chapter 3 | Lesson 9

117

Think and Grow: Use Properties to Multiply

You can use properties to multiply.

Example Find 8×250.

$8 \times 250 = (4 \times 2) \times 250$ Think: $8 = 4 \times 2$

$ = 4 \times (2 \times 250)$ Associative Property of Multiplication

$ = 4 \times \underline{}$

$ = \underline{}$ So, $8 \times 250 = \underline{}$.

Remember, you can use the Distributive Property with subtraction.

Example Find 5×698.

$5 \times 698 = 5 \times (700 - 2)$ Think: $698 = 700 - 2$

$ = (5 \times 700) - (5 \times 2)$ Distributive Property

$ = \underline{} - \underline{}$

$ = \underline{}$ So, $5 \times 698 = \underline{}$.

Example Find $4 \times 9 \times 25$.

$4 \times 9 \times 25 = 9 \times 4 \times 25$ Commutative Property of Multiplication

$ = 9 \times \underline{}$

$ = \underline{}$ So, $4 \times 9 \times 25 = \underline{}$.

Show and Grow I can do it!

Use properties to find the product. Explain your reasoning.

1. 6×150

2. 3×494

3. $25 \times 7 \times 4$

Apply and Grow: Practice

Use properties to find the product. Explain your reasoning.

4. 7×798

5. 350×6

6. 106×5

7. 4×625

8. 395×8

9. $2 \times 7 \times 15$

10. 430×2

11. 8×150

12. $3 \times 1,997$

13. $25 \times 9 \times 2$

14. 404×6

15. $4 \times 2,004$

16. Which One Doesn't Belong? Which expression does *not* belong with the other three?

$(3 \times 30) + (3 \times 7)$ $(3 \times 40) - (3 \times 3)$

$3 \times (30 + 7)$ $3 \times 3 \times 7$

17. Number Sense Use properties to find each product.

$9 \times 80 = 720$, so $18 \times 40 =$ _____.

$5 \times 70 = 350$, so $5 \times 72 =$ _____.

Chapter 3 | Lesson 9   119

Think and Grow: Modeling Real Life

Example   The fastest recorded speed of a dragster car in the United States was 31 miles per hour less than 3 times the top speed of the roller coaster. What was the speed of the car?

Multiply to find 3 times the top speed of the roller coaster.

$3 \times 120 = 3 \times (100 + 20)$ Rewrite 120 as 100 + 20.

$ = (3 \times 100) + (3 \times 20)$ Distributive Property

$ = \underline{} + \underline{}$

$ = \underline{}$ miles per hour

Top speed: 120 miles per hour

Subtract to find 31 miles per hour less.

$\underline{} - 31 = \underline{}$ miles per hour

So, the speed of the car was _____ miles per hour.

Show and Grow   I can think deeper!

18. In 2016, a theme park used 300 drones for a holiday show. In 2017, China used 200 fewer than 4 times as many drones for a lantern festival. How many drones did China use?

19. A subway train has 8 cars. Each car can hold 198 passengers. How many passengers can two subway trains hold?

20. You plant cucumbers, green beans, squash, and corn in a community garden. You plant 3 rows of each vegetable with 24 seeds in each row. How many seeds do you plant?

Name _____

Learning Target: Use properties to multiply.

Homework & Practice 3.9

You can use properties to multiply.

Example Find 7 × 496.

$7 \times 496 = 7 \times (500 - 4)$ Think: 496 = 500 − 4

$= (7 \times 500) - (7 \times 4)$ Distributive Property

$= \underline{3{,}500} - \underline{28}$

$= \underline{3{,}472}$

So, 7 × 496 = **3,472**.

Example Find 2 × 5 × 25.

$2 \times 5 \times 25 = 5 \times 2 \times 25$ Commutative Property of Multiplication

$= 5 \times \underline{50}$

$= \underline{250}$

So, 2 × 5 × 25 = **250**.

Use properties to find the product. Explain your reasoning.

1. 3 × 497

2. 36 × 9

3. 8 × 350

4. 25 × 8 × 4

5. 999 × 5

6. 9 × 402

7. 509 × 4

8. 2 × 9 × 15

9. 3,998 × 7

Chapter 3 | Lesson 9   121

10. YOU BE THE TEACHER Is Descartes correct? Explain.

$$895 \times 4 = (900 - 5) \times 4$$
$$= (900 \times 4) + (5 \times 4)$$
$$= 3,600 + 20$$
$$= 3,620$$

11. DIG DEEPER! Complete the square so that the product of each row and each column is 2,400.

4		
200		2
		2

12. Modeling Real Life The height of the Great Pyramid of Giza is 275 feet shorter than 2 times the height of the Luxor Hotel in Las Vegas. How tall is the Great Pyramid of Giza?

365 ft

Luxor Hotel in Las Vegas

13. Modeling Real Life Some firefighters are testing their equipment. Water from their firetruck hose hits the wall 12 feet higher than 7 times where the spray from a fire extinguisher hits the wall. How many feet high does the water from the hose hit the wall?

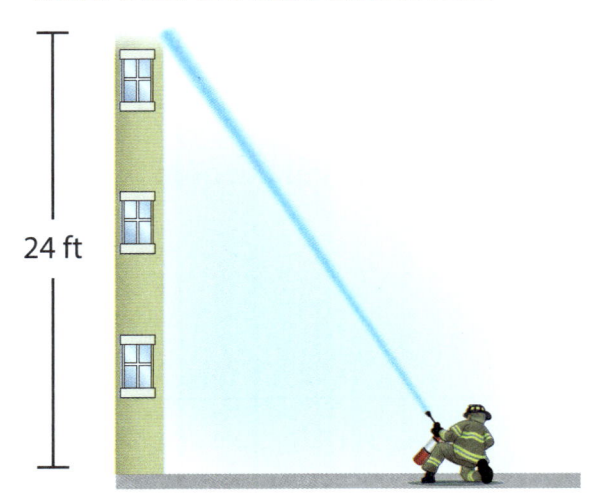

24 ft

Review & Refresh

14. You want to learn 95 new vocabulary words. You learn 5 words the first week and an equal number of words each week for the next 9 weeks. How many words do you learn in each of the 9 weeks?

Name _____

Problem Solving: Multiplication 3.10

Learning Target: Solve multi-step word problems involving multiplication.

Success Criteria:
- I can understand a problem.
- I can make a plan to solve using letters to represent the unknown numbers.
- I can solve a problem using an equation.

Explore and Grow

Make a plan to solve the problem.

A group of sea otters that swim together is called a raft. There are 37 otters in a raft. Each otter eats about 16 pounds of food each day. About how many pounds of food does the raft eat in 1 week?

MP **Critique Reasoning** Compare your plan to your partner's. How are your plans alike? How are they different?

Chapter 3 | Lesson 10   123

Think and Grow: Problem Solving: Multiplication

Example A coach buys 6 cases of sports drinks and spends $60. Each case has 28 bottles. A team drinks 85 bottles at a tournament. How many bottles are left?

Understand the Problem

What do you know?
- The coach buys 6 cases.
- The coach spends $60.
- Each case has 28 bottles.
- The team drinks 85 bottles.

What do you need to find?
- You need to find how many bottles are left.

Make a Plan

How will you solve?
- Multiply 28 by 6 to find the total number of bottles in 6 cases.
- Then subtract 85 bottles from the product to find how many bottles are left.
- The amount of money the coach spends is unnecessary information.

Solve

Step 1: How many bottles are in 6 cases?

| 28 | 28 | 28 | 28 | 28 | 28 |

⊢———— k ————⊣

k is the unknown product.

$$28 \times 6 = k$$

$$\begin{array}{r} 2\ 8 \\ \times 6 \\ \hline \end{array}$$

$k =$ _____

Step 2: Use k to find how many bottles are left.

| 85 | n |

⊢—— k = ____ ——⊣

n is the unknown difference.

$$\underline{} - 85 = n$$

$$\begin{array}{r} \\ -85 \\ \hline \end{array}$$

$n =$ _____

There are _____ bottles left.

Show and Grow   I can do it!

1. Explain how you can check whether your answer above is reasonable.

124

Name _____

Homework & Practice 3.10

Learning Target: Solve multi-step word problems involving multiplication.

Example You earn 206 points in Round 1 of a game. In Round 2, you earn 1,341 points. Then you find a magic star that makes your points from Round 2 worth 7 times as much. How many points do you earn in all?

Think: What do you know? What do you need to find? How will you solve?

Step 1: How many points do you earn from having star power?

| 1,341 | 1,341 | 1,341 | 1,341 | 1,341 | 1,341 | 1,341 |

⟵——— p ———⟶

p is the unknown product.

$1{,}341 \times 7 = p$

```
  2 2
  1, 3 4 1
×         7
─────────────
  9, 3 8 7
```

$p = \underline{9{,}387}$

Step 2: Use *p* to find how many points you earn in all.

206 | p = 9,387

⟵——— t ———⟶

t is the unknown sum.

$206 + \underline{9{,}387} = t$

```
    206
+ 9,387
───────
  9,593
```

$t = 9{,}593$

You earn __9,593__ points in all.

Understand the problem. Then make a plan. How will you solve? Explain.

1. *Raku* is a Japanese-inspired art form. An artist *fires*, or bakes, a raku pot at 1,409 degrees Fahrenheit. The artist fires a porcelain pot at 266 degrees Fahrenheit less than 2 times the temperature at which the raku pot is fired. You want to find the temperature at which the porcelain pot is fired.

2. You practice basketball 6 times each week. Each basketball practice is 55 minutes long. You practice dance for 225 minutes altogether each week. You want to find how many minutes you practice basketball and dance in all each week.

Chapter 3 | Lesson 10

3. You buy 7 books of stamps. There are 35 stamps in each book. You give some away and have 124 stamps left. How many stamps did you give away?

4. Your neighbor fills his car's gasoline tank with 9 gallons of gasoline. Each gallon of gasoline allows him to drive 23 miles. Can he drive for 210 miles without filling his gasoline tank? Explain.

5. Writing Write and solve a two-step word problem that can be solved using multiplication as one step.

6. Modeling Real Life There are 1,203 pictures taken for a yearbook. There are 124 student pictures for each of the 6 grades. There are also 7 pictures for each of the 23 school clubs. The rest of the pictures are teacher or candid pictures. How many teacher or candid pictures are there?

7. Modeling Real Life A construction worker earns $19 each hour she works. A supervisor earns $35 each hour she works. How much money do the construction worker and supervisor earn in all after 8 hours of work?

8. Modeling Real Life A family of 8 has $1,934 to spend on a vacation that is 1,305 miles away. They buy a $217 plane ticket and an $8 shirt for each person. How much money does the family have left?

Review & Refresh

What fraction of the whole is shaded?

9. 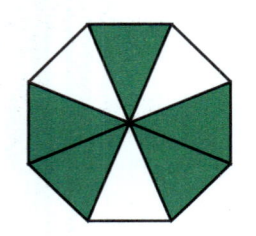 $\frac{\square}{\square}$ is shaded.

10. 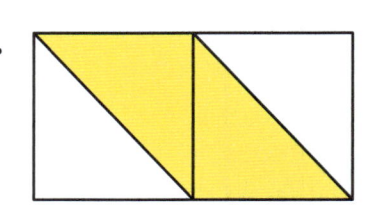 $\frac{\square}{\square}$ is shaded.

Name _____

Performance Task 3

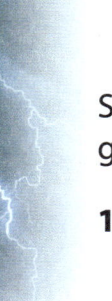

Sounds are vibrations that travel as waves through solids, liquids, and gases. Sound waves travel 1,125 feet per second through air.

1. You see a flash of lightning 5 seconds before you hear the thunder. How far away is the storm?

_____

2. Sound waves travel 22,572 feet per second faster through iron than through diamond. The speed of sound through diamond is 39,370 feet per second.
   a. Estimate the speed of sound through iron in feet per second.

   _____

   b. What is the actual speed of sound through iron in feet per second?

   _____

   c. Is your estimate close to the exact speed of sound through iron? Explain.

_____

3. Sound waves travel about 4 times faster through water than through air.
   a. What is the speed of sound through water in feet per second?

   _____

   b. A horn blows under water. A diver is about 9,000 feet away from the horn. About how many seconds does it take the diver to hear the sound of the horn?

_____

4. Do sound waves travel the fastest through solids, liquids, or gases? Explain.

Chapter 3

Multiplication Quest

Directions:

1. Players take turns rolling a die. Players solve problems on their boards to race the knights to their castles.
2. On your turn, solve the next multiplication problem in the row of your roll.
3. The first player to get a knight to a castle wins!

Die	Knight					Castle
⚀		7 × 8	32 × 3	629 × 4	5,107 × 6	
⚁		3 × 9	56 × 4	248 × 1	3,816 × 8	
⚂		5 × 2	81 × 5	921 × 9	7,249 × 7	
⚃		4 × 6	90 × 9	455 × 8	9,683 × 2	
⚄		8 × 1	12 × 7	806 × 3	4,749 × 5	
⚅		6 × 7	79 × 2	573 × 6	8,106 × 4	

Name _____

Chapter Practice 3

3.1 Understand Multiplicative Comparisons

Write two comparison sentences for the equation.

1. $72 = 8 \times 9$

2. $60 = 10 \times 6$

Write an equation for the comparison sentence.

3. 28 is 4 times as many as 7.

4. 40 is 8 times as many as 5.

5. Newton saves $40. Descartes saves $25 more than Newton. How much money does Descartes save?

6. Your friend is 10 years old. Your neighbor is 4 times as old as your friend. How old is your neighbor?

3.2 Multiply Tens, Hundreds, and Thousands

Find the product.

7. $6 \times 90 =$ _____

8. $3{,}000 \times 1 =$ _____

9. $4 \times 200 =$ _____

10. $4{,}000 \times 4 =$ _____

11. $8 \times 700 =$ _____

12. $2 \times 60 =$ _____

Find the missing factor.

13. _____ $\times 200 = 1{,}000$

14. $5 \times$ _____ $= 450$

15. _____ $\times 800 = 6{,}400$

Chapter 3   131

3.3 Estimate Products by Rounding

Estimate the product.

16. 5 × 65

17. 2 × 903

18. 7 × 3,592

Find two estimates that the product is between.

19. 8 × 32

20. 4 × 284

21. 6 × 5,945

22. A charity organizer raises $9,154 each month for 6 months. To determine whether the charity raises $50,000, can you use an estimate, or is an exact answer required? Explain.

3.4 Use Distributive Property to Multiply

Find the product.

23. 3 × 14 = _____

24. 18 × 9 = _____

25. 36 × 5 = _____

26. 8 × 56 = _____

27. 6 × 67 = _____

28. 83 × 2 = _____

29. **Structure** Use the Distributive Property to write an equation shown by the model.

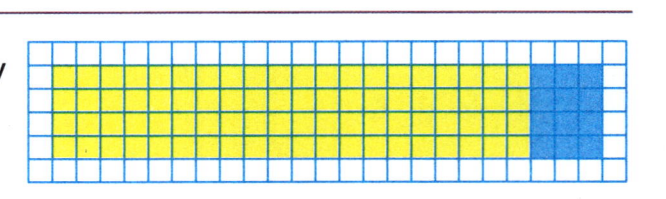

132

3.5 Use Expanded Form to Multiply

Find the product.

30. 487 × 3 = _____

31. 5 × 7,402 = _____

32. 8,395 × 7 = _____

3.6 Use Partial Products to Multiply

Find the product.

33. 266
      ×   9

34.   85
      ×   8

35. 7,032
      ×   4

36. **Number Sense** Which three numbers are the partial products that you add to find the product of 518 and 2?

      100   200   16   1,000   14   20

3.7 Multiply Two-Digit Numbers by One-Digit Numbers

Find the product. Check whether your answer is reasonable.

37. Estimate: _____

      19
  × 6

38. Estimate: _____

      73
  × 7

39. Estimate: _____

      58
  × 5

Chapter 3

3.8 Multiply Three- and Four-Digit Numbers by One-Digit Numbers

Find the product. Check whether your answer is reasonable.

40. Estimate: _____

402 × 3 = _____

41. Estimate: _____

8 × 3,861 = _____

42. Estimate: _____

977 × 2 = _____

Compare.

43. 308 × 6 ◯ 5 × 408

44. 6 × 789 ◯ 2 × 2,367

45. 454 × 3 ◯ 313 × 4

3.9 Use Properties to Multiply

Use properties to find the product. Explain your reasoning.

46. 25 × 9 × 4

47. 8 × 250

48. 3 × 497

49. 2 × 8 × 15

50. 699 × 9

51. 1,003 × 6

3.10 Problem Solving: Multiplication

52. A musical cast sells 1,761 tickets for a big show. The cast needs to complete 36 days of rehearsal. Each rehearsal is 8 hours long. The cast has rehearsed for 102 hours so far. How many hours does the cast have left to rehearse?

Name _____

Cumulative Practice 1-3

1. Which number is greater than 884,592?

   Ⓐ 89,621 Ⓑ 884,592

   Ⓒ 805,592 Ⓓ 894,592

2. What is the difference of 30,501 and 6,874?

3. A teenager will send about 37,000 text messages within 1 year. Which numbers could be the exact number of text messages sent?

   ☐ 37,461   ☐ 36,834   ☐ 37,050

   ☐ 37,503   ☐ 36,006   ☐ 36,752

4. What is the product of 4,582 and 6?

   Ⓐ 27,492 Ⓑ 24,082

   Ⓒ 27,432 Ⓓ 42,117

5. Which expression is equal to 246,951 + 73,084?

   Ⓐ 246,951 + 70,000 + 3,000 + 800 + 40   Ⓑ 246,951 + 7,000 + 300 + 80 + 4

   Ⓒ 246,951 + 70,000 + 3,000 + 80 + 4   Ⓓ 246,951 + 70,000 + 3,000 + 800 + 4

Chapter 3   135

6. Newton reads the number "four hundred six thousand, twenty-nine" in a book. What is this number written in standard form?

   Ⓐ 4,629
   Ⓑ 406,029
   Ⓒ 460,029
   Ⓓ 406,290

7. Which expressions have a product of 225?

   ☐ 9 × 25
   ☐ 75 × 3
   ☐ 56 × 4
   ☐ 5 × 45

8. What is the greatest possible number you can make with the number cards below?

   [6]   [1]   [8]   [4]   [2]

   Ⓐ 84,621
   Ⓑ 86,421
   Ⓒ 12,468
   Ⓓ 68,421

9. What is the sum of 62,671 and 48,396?

   Ⓐ 111,067
   Ⓑ 11,067
   Ⓒ 100,967
   Ⓓ 100,067

10. Your friend drinks 8 glasses of water each day. He wants to know how many glasses of water he will drink in 1 year. Between which two estimates is the number of glasses he will drink in a year?

    Ⓐ 2,400 and 3,200
    Ⓑ 240 and 320
    Ⓒ 3,200 and 4,000
    Ⓓ 320 and 400

11. Which statement about the number 420,933 is true?

- (A) The value of the 2 is 2,000.
- (B) The 4 is in the ten thousands place.
- (C) The value of the 3 in the tens place is ten times the value of the 3 in the ones place.
- (D) There are 9 tens.

12. Descartes rounds to the nearest ten thousand and gets an estimate of 360,000. Which expressions could be the problem he estimated?

- ☐ 785,462 − 432,587
- ☐ 480,012 − 127,465
- ☐ 74,621 − 10,354
- ☐ 398,650 − 41,579

13. There are 8 students in a book club. There are 5 times as many students in a drama club as the book club. How many students are in the drama club?

- (A) 13 students
- (B) 45 students
- (C) 3 students
- (D) 40 students

14. The force required to shatter concrete is 3 times the amount of force required to shatter plexiglass. The force required to shatter plexiglass is 72 pounds per square foot. What is the force required to shatter concrete?

- (A) 216 pounds per square foot
- (B) 69 pounds per square foot
- (C) 75 pounds per square foot
- (D) 2,106 pounds per square foot

15. Which number, when rounded to the nearest hundred thousand, is equal to 100,000?

- (A) 9,802
- (B) 83,016
- (C) 152,853
- (D) 46,921

Chapter 3

16. You want to find 5,193 × 8. Which expressions show how to use the Distributive Property to find the product?

   ☐ 8 × (5,000 + 100 + 90 + 3)
   ☐ (8 × 5,000) + (8 × 100) + (8 × 90) + (8 × 3)
   ☐ 40,000 + 800 + 720 + 24
   ☐ (8 × 3) + (8 × 9) + (8 × 1) + (8 × 5)

17. Use the table to answer the questions.

   Think Solve Explain

   Part A Five friends each buy all of the items listed in the table. Use rounding to find about how much money they spend in all.

   Trampoline Park Prices

Item	Price
90-minute jump pass	$21
Water Bottle	$8
T-shirt	$15

   Part B An Ultimate Package costs $40 and includes all of the items listed in the table. About how much money would the 5 friends have saved in all if they would have bought an Ultimate Package instead of purchasing the items individually? Explain.

18. A pair of walruses weigh 4,710 pounds together. The female weighs 1,942 pounds. How much more does the male weigh than the female?

   Ⓐ 826 pounds
   Ⓑ 6,652 pounds
   Ⓒ 2,768 pounds
   Ⓓ 2,710 pounds

19. You use compensation as shown. What is the final step?

   ```
     6,325              6,325
   − 2,258  − 58  →   − 2,200
                        4,125
   ```

   Ⓐ Add 58 to 4,125.
   Ⓑ Subtract 58 from 4,125.
   Ⓒ Subtract 4,125 from 6,325.
   Ⓓ There is no final step. You are finished.

138

Name _____

STEAM Performance Task 1-3

The Golden Gate Bridge

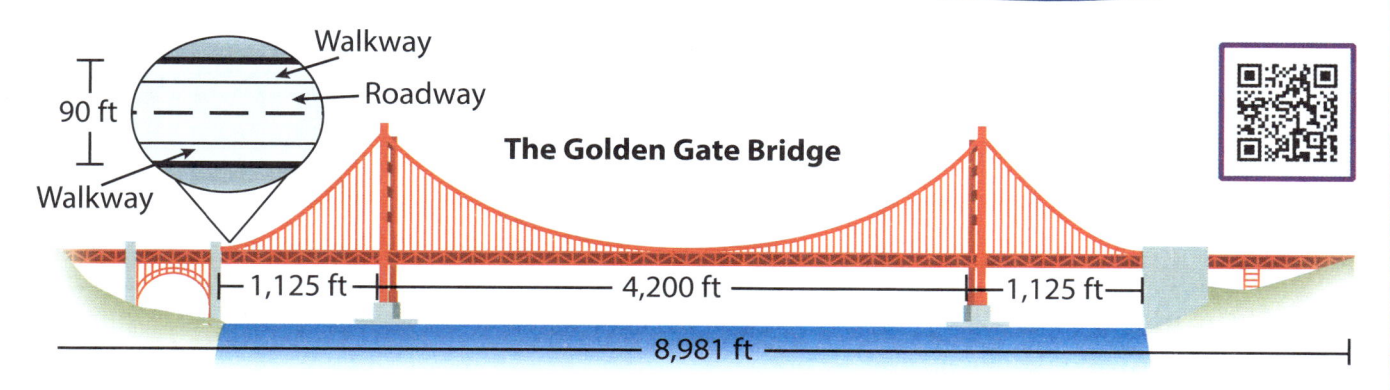

- 90 ft (Walkway, Roadway, Walkway)
- 1,125 ft — 4,200 ft — 1,125 ft
- 8,981 ft

A teacher visits the Golden Gate Bridge in San Francisco, California.

1. Use the figure above to answer the questions.

   a. The roadway is 8 times as wide as width of both walkways combined. Each walkway has the same width. How wide is one of the walkways?

   b. What is the length of the bridge that is suspended above the water?

   c. What is the length of the bridge that is *not* suspended above the water?

2. The teacher wants to estimate the distance between the bridge and the water.

   a. He rides in a ferry boat under the bridge. He says that the distance between the bridge and the water is about 5 times the height of the ferry boat. What is the distance between the bridge and the water?

   44 ft

   b. The teacher estimates that the height of one tower is about three times the distance between the bridge and the water. What is the teacher's estimate for the height of the tower?

   c. You learn the exact height of the tower is 86 feet taller than the teacher's estimate. How tall is the tower?

Chapter 3

The teacher uses his fitness tracker to count the number of steps he walks on the Golden Gate Bridge each day for 1 week.

Day	Number of Steps
Sunday	12,378
Monday	10,712
Tuesday	7,989
Wednesday	
Thursday	
Friday	
Saturday	

3. The table shows the numbers of steps taken on the first 3 days of the week. Use this information to complete the table.

   a. He takes one thousand one hundred twenty-two more steps on Wednesday than on Monday.

   b. On Thursday, he takes one hundred eighteen steps less than on Sunday.

   c. He takes ten thousand fifty steps on Friday.

   d. On Saturday, he takes twice as many steps as on Tuesday.

4. The teacher's goal is to take 11,500 steps each day.

   a. What is his goal for the week?

   b. Estimate the total number of steps he takes from Sunday to Saturday. Does he meet his goal for the week? Explain.

   c. One mile is about two thousand steps. The teacher estimates that he walks 3 miles from one end of the Golden Gate Bridge to the other and back again. About how many steps does he walk?

140

4 Multiply by Two-Digit Numbers

- Wind turbines convert wind into energy. The wind is a renewable source of energy. What are some other forms of renewable energy?

- The amount of energy generated is based on the speed of the wind and the lengths of the turbine blades. What effect do you think the blade lengths have on the amount of energy a wind turbine can generate?

Chapter Learning Target:
Understand multiplying two-digit numbers.

Chapter Success Criteria:
- I can find the product of two numbers.
- I can use rounding to estimate a product.
- I can write multiplication problems.
- I can solve a problem using an equation.

4 Vocabulary

Name _____

Review Words
place value
thousands

Organize It

Use the review words to complete the graphic organizer.

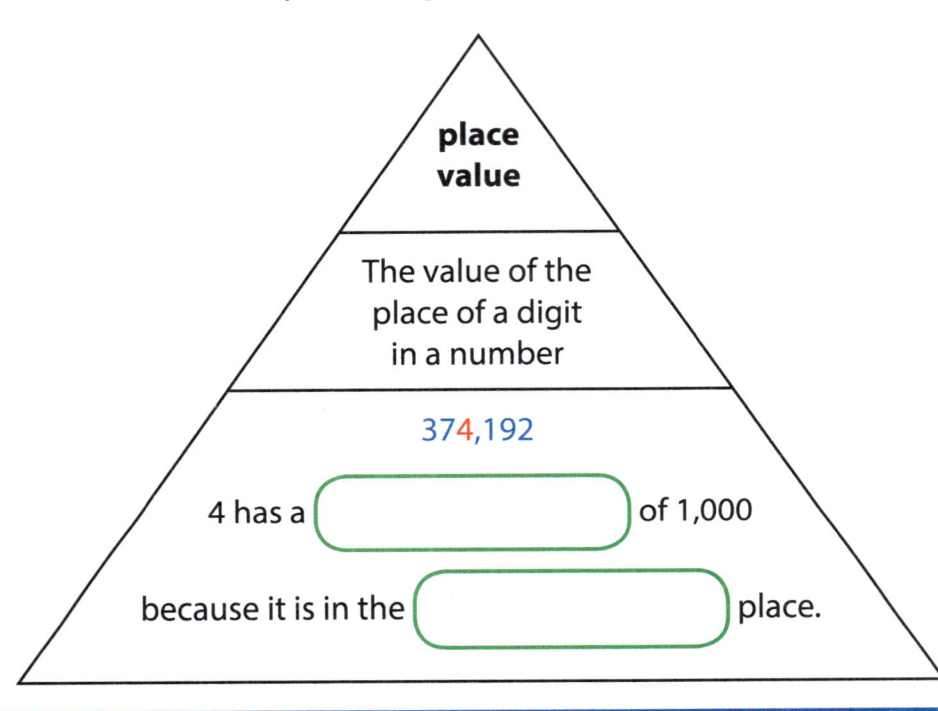

place value

The value of the place of a digit in a number

374,192

4 has a [_____] of 1,000

because it is in the [_____] place.

Define It

What am I?

Numbers that are easy to multiply and are close to the actual numbers

3 × 7 = T   6 × 5 = A   5 × 9 = C   3 × 8 = M   4 × 5 = O
9 × 2 = B   8 × 9 = N   7 × 6 = P   8 × 6 = I   7 × 10 = L
7 × 4 = R   4 × 2 = E   10 × 5 = U   6 × 9 = S

45	20	24	42	30	21	48	18	70	8

72	50	24	18	8	28	54

142

Chapter 4 Vocabulary Cards

compatible numbers

Numbers that are easy to multiply and are close to the actual numbers

$$24 \times 31$$
$$\downarrow \quad \downarrow$$
$$25 \times 30$$

Name _____

Multiply by Tens 4.1

Learning Target: Use place value and properties to multiply by multiples of ten.

Success Criteria:
- I can use place value to multiply by multiples of ten.
- I can use the Associative Property to multiply by multiples of ten.
- I can describe a pattern with zeros when multiplying by multiples of ten.

Explore and Grow

Model each product. Draw each model.

$2 \times 3 = $ _____	$2 \times 30 = $ _____
$2 \times 300 = $ _____	$2 \times 3{,}000 = $ _____

What pattern do you notice in the products?

MP Repeated Reasoning How can the pattern above help you find 20×30?

Chapter 4 | Lesson 1   143

Think and Grow: Multiply by Multiples of Tens

You can use place value and properties to multiply two-digit numbers by multiples of ten.

Example Find 40 × 20.

One Way: Use place value.

40 × 20 = 40 × _____ tens

= _____ tens

= _____

So, 40 × 20 = _____.

Example Find 12 × 30.

One Way: Use place value.

You can use regrouping, a number line, or partial products to find 12 × 3.

12 × 30 = 12 × _____ tens

= _____ tens

= _____

So, 12 × 30 = _____.

Another Way: Use the Associative Property of Multiplication.

40 × 20 = 40 × (2 × 10)   Rewrite 20 as 2 × 10.

= (40 × 2) × 10   Associative Property of Multiplication

= _____ × 10

= _____

So, 40 × 20 = _____.

Another Way: Use the Associative Property of Multiplication.

12 × 30 = 12 × (3 × 10)   Rewrite 30 as 3 × 10.

= (12 × 3) × 10   Associative Property of Multiplication

= _____ × 10

= _____

So, 12 × 30 = _____.

Show and Grow   I can do it!

Find the product.

1. 70 × 40 = _____

2. 50 × 80 = _____

3. 24 × 90 = _____

4. 45 × 60 = _____

Name _____

Estimate Products 4.2

Learning Target: Use rounding and compatible numbers to estimate products.

Success Criteria:
- I can use rounding to estimate a product.
- I can use compatible numbers to estimate a product.
- I can explain different ways to estimate a product.

Explore and Grow

Choose an expression to estimate each product. Write the expression. You may use an expression more than once.

| 20 × 20 | 20 × 25 |
| 25 × 40 | 40 × 20 |

21 × 24   26 × 38   23 × 17   42 × 23

___ × ___   ___ × ___   ___ × ___   ___ × ___

Compare your answers with a partner. Did you choose the same expressions?

MP Construct Arguments Which estimated product do you think will be closer to the product of 29 and 37? Explain your reasoning.

25 × 40

30 × 40

Chapter 4 | Lesson 2   149

Think and Grow: Estimate Products

You can estimate products using rounding or compatible numbers. **Compatible numbers** are numbers that are easy to multiply and are close to the actual numbers.

Example Use rounding to estimate 57 × 38.

Step 1: Round each factor to the nearest ten.

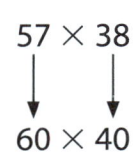

57 × 38
↓   ↓
60 × 40

Step 2: Multiply.

60 × 40 = 60 × _____ tens

= _____ tens

= _____

So, 57 × 38 is about _____.

Example Use compatible numbers to estimate 24 × 31.

Step 1: Choose compatible numbers.

Think: 24 is close to 25.
31 is close to 30.

24 × 31
↓   ↓
25 × 30

Step 2: Multiply.

25 × 30 = 25 × _____ tens

= _____ tens

= _____

So, 24 × 31 is about _____.

Show and Grow I can do it!

Use rounding to estimate the product.

1. 27 × 50

2. 42 × 14

3. 61 × 73

Use compatible numbers to estimate the product.

4. 19 × 26

5. 23 × 78

6. 74 × 20

Name _____

Homework & Practice 4.2

Learning Target: Use rounding and compatible numbers to estimate products.

Example Use rounding to estimate 45 × 43.

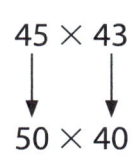

Step 1: Round each factor to the nearest ten.

45 × 43
↓   ↓
50 × 40

Step 2: Multiply.

50 × 40 = 50 × __4__ tens

= __200__ tens

= __2,000__

So, 45 × 43 is about __2,000__.

Example Use compatible numbers to estimate 61 × 24.

Step 1: Choose compatible numbers.

Think: 61 is close to 60.
24 is close to 25.

61 × 24
↓   ↓
60 × 25

Step 2: Multiply.

60 × 25 = __6__ tens × 25

= __150__ tens

= __1,500__

So, 61 × 24 is about __1,500__.

Use rounding to estimate the product.

1. 42 × 13

2. 56 × 59

3. 19 × 91

Use compatible numbers to estimate the product.

4. 23 × 78

5. 67 × 45

6. 19 × 24

Chapter 4 | Lesson 2

Estimate the product.

7. 84 × 78

8. 92 × 34

9. 57 × 81

Open-Ended Write two possible factors that could be estimated as shown.

10. 6,400

_____ × _____

↓   ↓

_____ × _____ = 6,400

11. 1,600

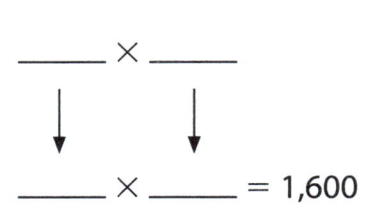

_____ × _____

↓   ↓

_____ × _____ = 1,600

12. Reasoning Are both Newton's and Descartes's estimates reasonable? Explain.

27 × 68 = 30 × 70
       = 2,100

27 × 68 = 25 × 70
       = 1,750

13. DIG DEEPER! You use 90 × 30 to estimate 92 × 34. Will your estimate be greater than or less than the actual product? Explain.

14. Modeling Real Life About how many hours of darkness does Barrow, Alaska have in December?

Remember, there are 24 hours in 1 day.

Days of Darkness in Barrow, Alaska

Month	Days
November	12
December	31
January	21

Review & Refresh

15. Round 253,490 to the nearest ten thousand.

16. Round 628,496 to the nearest hundred thousand.

Name _____

Use Area Models to Multiply Two-Digit Numbers 4.3

Learning Target: Use area models and partial products to multiply.

Success Criteria:
- I can use an area model to break apart the factors of a product.
- I can relate an area model to partial products.
- I can add partial products to find a product.

Explore and Grow

Draw an area model that represents 15 × 18. Then break apart your model into smaller rectangles.

What is the total area of your model? Explain how you found your answer.

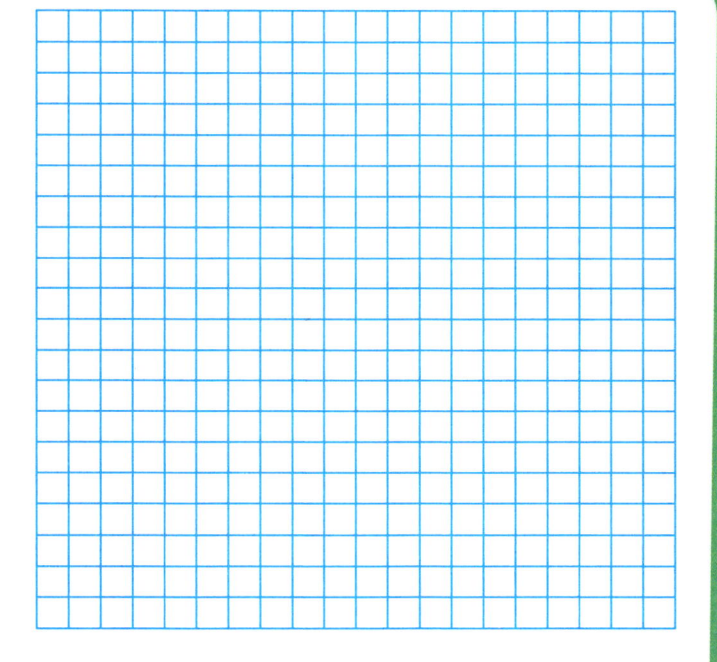

MP Reasoning Compare with a partner. Do you get the same answer? Explain.

Chapter 4 | Lesson 3

155

Think and Grow: Use Area Models to Multiply

Example Use an area model and partial products to find 12 × 14.

Model the expression. Break apart 12 as 10 + 2 and 14 as 10 + 4.

Why does the sum of the partial products represent the sum of the whole area?

Add the area of each rectangle to find the product for the whole model.

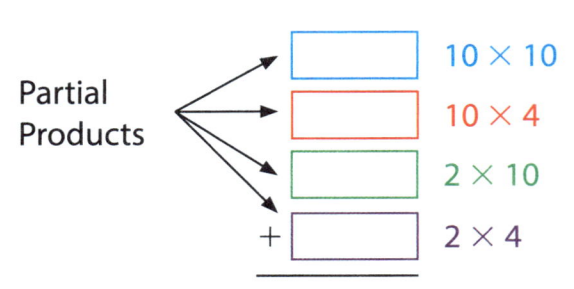

Partial Products
- ☐ 10 × 10
- ☐ 10 × 4
- ☐ 2 × 10
- + ☐ 2 × 4

_____ Add the partial products.

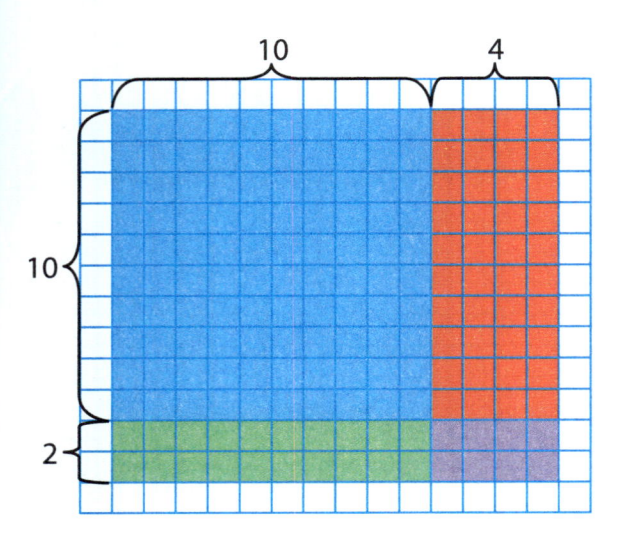

So, 12 × 14 = _____.

Show and Grow I can do it!

Use the area model to find the product.

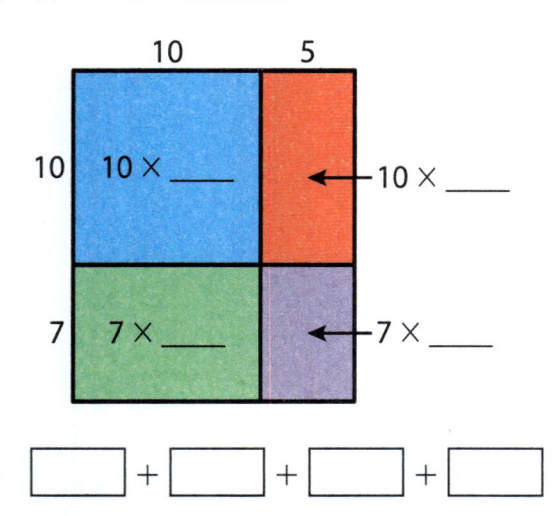

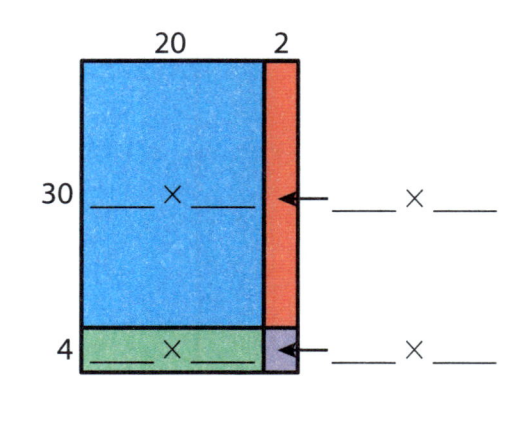

1. 17 × 15 = _____

10 × _____
10 × _____
7 × _____
7 × _____

☐ + ☐ + ☐ + ☐

2. 34 × 22 = _____

_____ × _____
_____ × _____
_____ × _____
_____ × _____

☐ + ☐ + ☐ + ☐

Name _____

Apply and Grow: Practice

Use the area model to find the product.

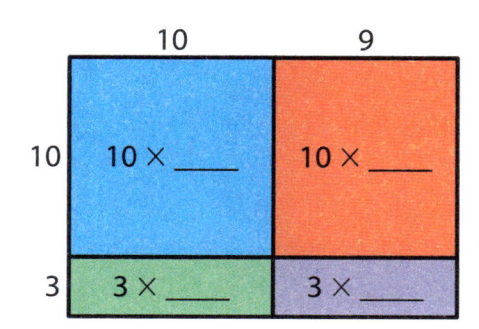

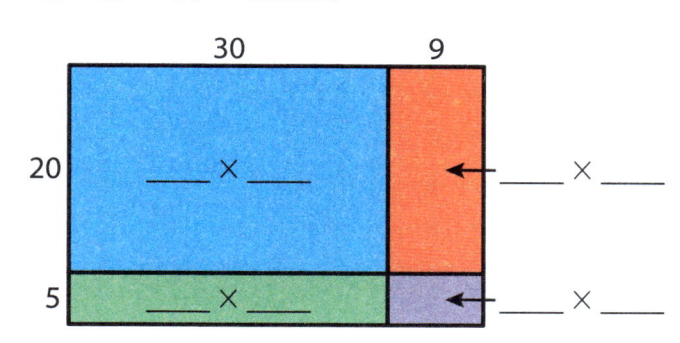

3. 13 × 19 = _____

	10	9
10	10 × ___	10 × ___
3	3 × ___	3 × ___

4. 25 × 39 = _____

	30	9
20	___ × ___	___ × ___
5	___ × ___	___ × ___

Draw an area model to find the product.

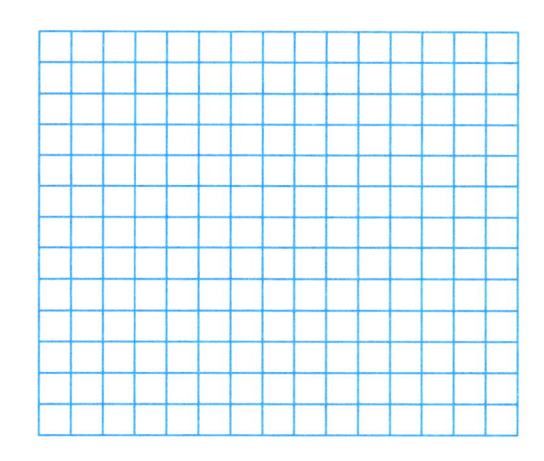

5. 11 × 13 = _____

6. 23 × 26 = _____

_____

7. 27 × 45 = _____

8. Perseid meteors travel 59 kilometers each second. How far does a perseid meteor travel in 15 seconds?

9. **DIG DEEPER!** Write the multiplication equation represented by the area model.

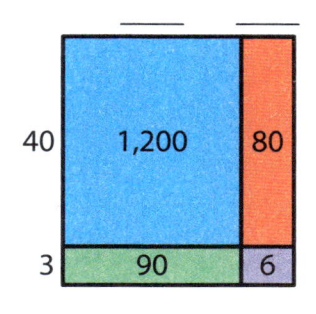

	___	___
40	1,200	80
3	90	6

_____ × _____ = _____

Chapter 4 | Lesson 3   157

Think and Grow: Modeling Real Life

Example A wind farm has 8 rows of new wind turbines and 3 rows of old wind turbines. Each row has 16 turbines. How many turbines does the wind farm have?

Add the number of rows of new turbines to the number of rows of old turbines.

$$8 + 3 = \underline{}$$

There are _____ rows of turbines.

Multiply the number of rows by the number in each row.

$$11 \times 16$$

	10 × 10
	10 × 6
	1 × 10
+	1 × 6

_____ Add the partial products.

The wind farm has _____ turbines.

Show and Grow   I can think deeper!

10. You can type 19 words per minute. Your cousin can type 33 words per minute. How many more words can your cousin type in 15 minutes than you?

11. A store owner buys 24 packs of solar eclipse glasses. Each pack has 12 glasses. The store did *not* sell 18 of the glasses. How many of the glasses did the store sell?

Name _____

Homework & Practice 4.3

Learning Target: Use area models and partial products to multiply.

Example Use an area model and partial products to find 15 × 18.

Model the expression. Break apart 15 as 10 + 5 and 18 as 10 + 8.

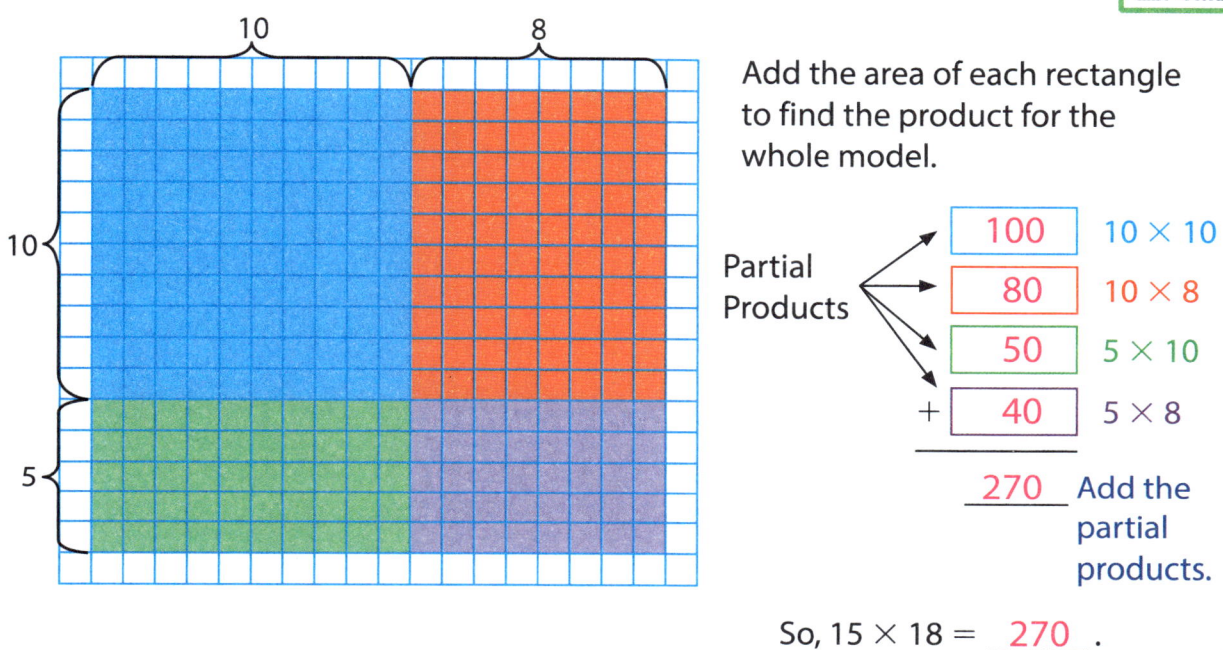

Add the area of each rectangle to find the product for the whole model.

Partial Products:
- 100   10 × 10
- 80   10 × 8
- 50   5 × 10
- + 40 5 × 8
- 270   Add the partial products.

So, 15 × 18 = __270__.

Use the area model to find the product.

1. 12 × 13 = _____

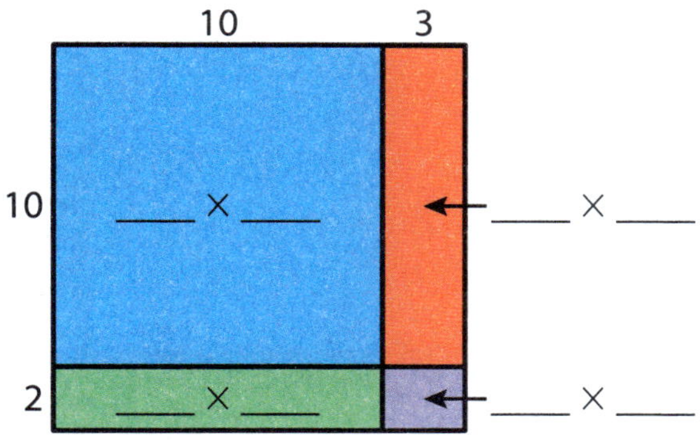

___ × ___   ___ × ___
___ × ___   ___ × ___

☐ + ☐ + ☐ + ☐

2. 38 × 24 = _____

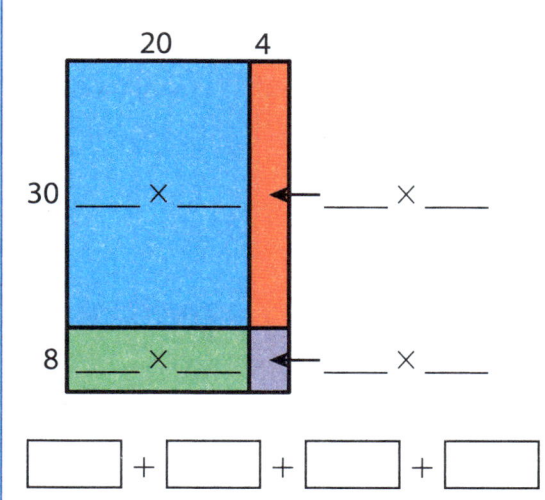

___ × ___   ___ × ___
___ × ___   ___ × ___

☐ + ☐ + ☐ + ☐

Chapter 4 | Lesson 3   159

Use the area model to find the product.

3. 19 × 18 = _____

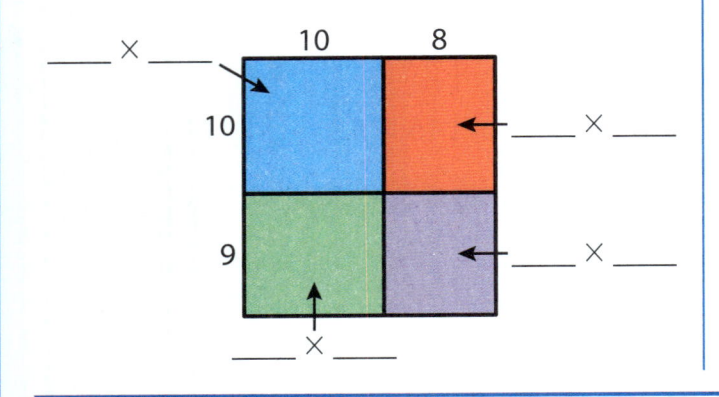

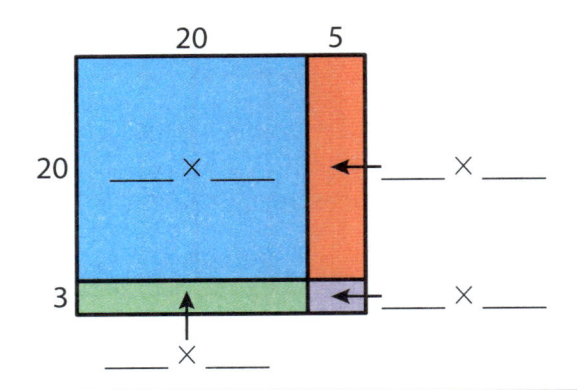

___ × ___

10 | 8
10
9
___ × ___
___ × ___
___ × ___

4. 23 × 25 = _____

20 | 5
20
3
___ × ___
___ × ___
___ × ___
___ × ___

Draw an area model to find the product.

5. 26 × 31 = _____

6. 22 × 47 = _____

7. YOU BE THE TEACHER Your friend finds 12 × 42. Is your friend correct? Explain.

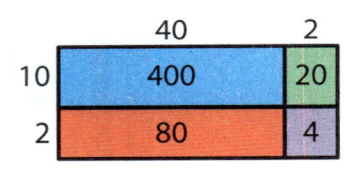

```
        40    2
   10  400   20
    2   80    4
```

400 + 80 + 20 + 4 = 504

8. Writing Explain how to use an area model and partial products to multiply two-digit numbers.

9. Modeling Real Life A mega-arcade has 9 rows of single-player games and 5 rows of multi-player games. Each row has 24 games. How many games does the arcade have?

Review & Refresh

Find the sum. Check whether your answer is reasonable.

10.   75,420
       +   8,596

11.   47,928
       + 23,657

12.   505,019
       + 64,802

Name _____

Use the Distributive Property to Multiply Two-Digit Numbers

4.4

Learning Target: Use area models and the Distributive Property to multiply.

Success Criteria:
- I can use an area model and partial products to multiply.
- I can use an area model and the Distributive Property to multiply.

Explore and Grow

Use as few base ten blocks as possible to create an area model for 13 × 24. Draw to show your model.

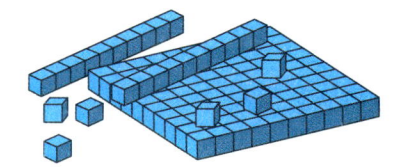

```
           24
    ┌─────────────────┐
    │                 │
 13 │                 │
    │                 │
    └─────────────────┘
```

Color your model to show four smaller rectangles. Label the partial products.

Reasoning How do you think the Distributive Property relates to your area model? Explain.

Chapter 4 | Lesson 4

161

Think and Grow: Use the Distributive Property to Multiply

Example Find 17×25.

One Way: Use an area model and partial products.

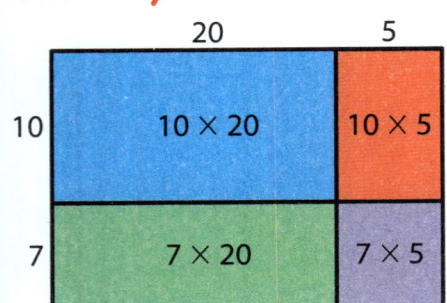

Add the partial products.

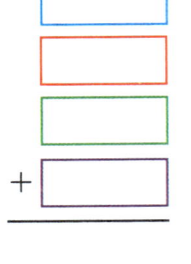

So, $17 \times 25 = $ _____.

Another Way: Use an area model and the Distributive Property.

$17 \times 25 = 17 \times (20 + 5)$ Break apart 25.

$= (17 \times 20) + (17 \times 5)$ Distributive Property

$= (10 + 7) \times 20 + (10 + 7) \times 5$ Break apart 17.

$= (10 \times 20) + (7 \times 20) + (10 \times 5) + (7 \times 5)$ Distributive Property

$= $ _____ $+$ _____ $+$ _____ $+$ _____

$= $ _____

So, $17 \times 25 = $ _____.

Show and Grow I can do it!

1. Use the area model and the Distributive Property to find 32×19.

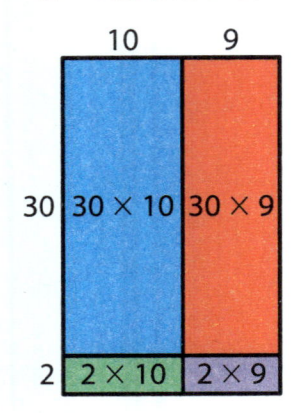

$32 \times 19 = 32 \times (10 + 9)$

$= (32 \times 10) + (32 \times 9)$

$= (30 + 2) \times $ _____ $+ (30 + 2) \times $ _____

$= (30 \times $ _____ $) + (2 \times $ _____ $) + (30 \times $ _____ $) + (2 \times $ _____ $)$

$= $ _____ $+$ _____ $+$ _____ $+$ _____

$= $ _____

So, $32 \times 19 = $ _____.

Name _____

Apply and Grow: Practice

2. Use the area model and the Distributive Property to find 34 × 26.

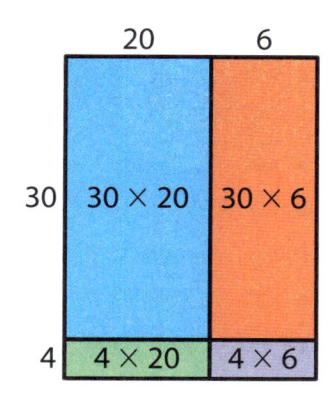

$34 \times 26 = 34 \times (20 + 6)$

$= (34 \times 20) + (34 \times 6)$

$= (30 + 4) \times \underline{} + (30 + 4) \times \underline{}$

$= (30 \times \underline{}) + (4 \times \underline{}) + (30 \times \underline{}) + (4 \times \underline{})$

$= \underline{} + \underline{} + \underline{} + \underline{}$

$= \underline{}$ So, $34 \times 26 = \underline{}$.

Use the Distributive Property to find the product.

3. $28 \times 47 = 28 \times (40 + 7)$

$= (28 \times 40) + (28 \times 7)$

$= (20 + 8) \times \underline{} + (20 + 8) \times \underline{}$

$= (20 \times \underline{}) + (8 \times \underline{}) + (20 \times \underline{}) + (8 \times \underline{})$

$= \underline{} + \underline{} + \underline{} + \underline{}$

$= \underline{}$ So, $28 \times 47 = \underline{}$.

4. $39 \times 41 = \underline{}$

5. $74 \times 12 = \underline{}$

6. $83 \times 65 = \underline{}$

7. Which One Doesn't Belong? Which expression does *not* belong with the other three?

$(40 + 7) \times 52$ $(40 + 7) \times (50 + 2)$ $(40 \times 7) \times (50 \times 2)$ $47 \times (50 + 2)$

Chapter 4 | Lesson 4

Think and Grow: Modeling Real Life

Example The dunk tank at a school fair needs 350 gallons of water. There are 27 students in a class. Each student pours 13 gallons of water into the tank. Is there enough water in the dunk tank?

Find how many gallons of water the students put in the dunk tank.

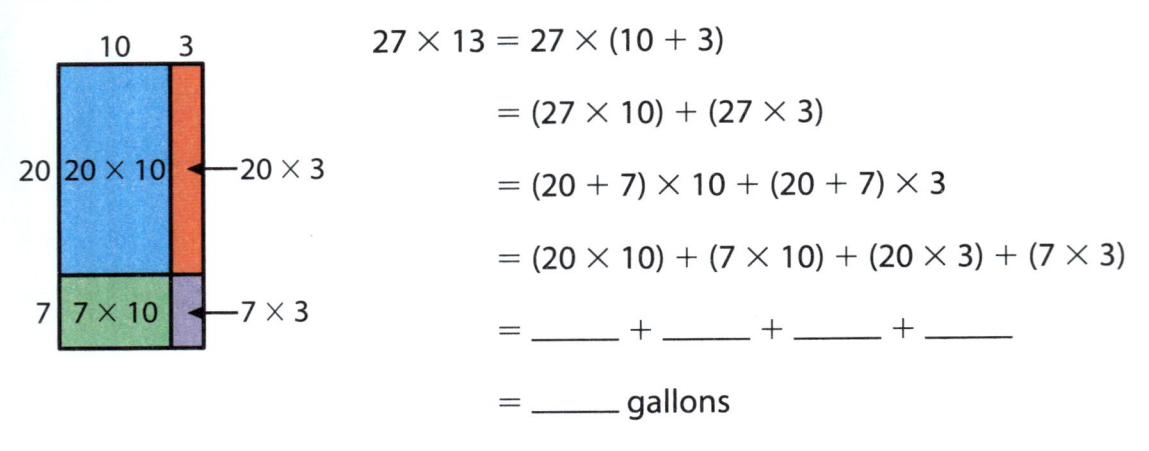

$27 \times 13 = 27 \times (10 + 3)$

$= (27 \times 10) + (27 \times 3)$

$= (20 + 7) \times 10 + (20 + 7) \times 3$

$= (20 \times 10) + (7 \times 10) + (20 \times 3) + (7 \times 3)$

$=$ _____ $+$ _____ $+$ _____ $+$ _____

$=$ _____ gallons

Compare the numbers of gallons.

So, there _____ enough water in the dunk tank.

Show and Grow   I can think deeper!

8. An event coordinator orders 35 boxes of T-shirts to give away at a baseball game. There are 48 T-shirts in each box. If 2,134 fans attend the game, will each fan get a T-shirt?

9. A horse owner must provide 4,046 square meters of pasture for each horse. Is the pasture large enough for 2 horses? Explain.

96 m

86 m

164

Name _____

Homework & Practice 4.4

Learning Target: Use area models and the Distributive Property to multiply.

Example Find 13 × 28.

One Way: Use an area model and partial products.

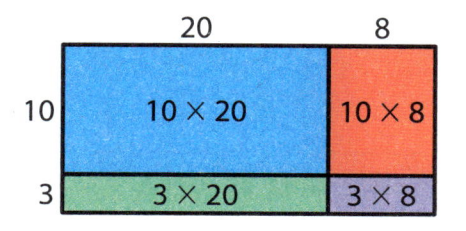

	20	8
10	10 × 20	10 × 8
3	3 × 20	3 × 8

Add the partial products.

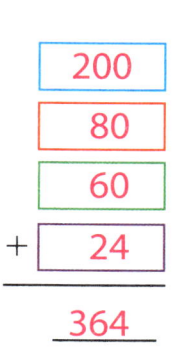

```
  200
   80
   60
+  24
-----
  364
```

So, 13 × 28 = __364__.

Another Way: Use an area model and the Distributive Property.

13 × 28 = 13 × (20 + 8)   Break apart 28.

= (13 × 20) + (13 × 8)   Distributive Property

= (10 + 3) × 20 + (10 + 3) × 8   Break apart 13.

= (10 × 20) + (3 × 20) + (10 × 8) + (3 × 8)   Distributive Property

= __200__ + __60__ + __80__ + __24__

= __364__   So, 13 × 28 = __364__.

1. Use the area model and the Distributive Property to find 45 × 21.

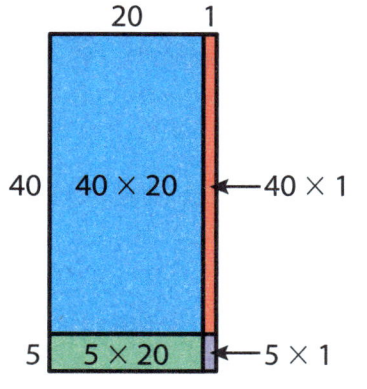

	20	1
40	40 × 20	40 × 1
5	5 × 20	5 × 1

45 × 21 = 45 × (20 + 1)

= (45 × 20) + (45 × 1)

= (40 + 5) × ____ + (40 + 5) × ____

= (40 × ____) + (5 × ____) + (40 × ____) + (5 × ____)

= ____ + ____ + ____ + ____

= ____   So, 45 × 21 = _____.

Chapter 4 | Lesson 4

2. Use the Distributive Property to find the product.

$34 \times 49 = 34 \times (40 + 9)$

$\qquad = (34 \times 40) + (34 \times 9)$

$\qquad = (30 + 4) \times \underline{\qquad} + (30 + 4) \times \underline{\qquad}$

$\qquad = (30 \times \underline{\qquad}) + (4 \times \underline{\qquad}) + (30 \times \underline{\qquad}) + (4 \times \underline{\qquad})$

$\qquad = \underline{\qquad} + \underline{\qquad} + \underline{\qquad} + \underline{\qquad}$

$\qquad = \underline{\qquad}$ So, $34 \times 49 = \underline{\qquad}$.

3. $14 \times 27 = \underline{\qquad}$

4. $38 \times 31 = \underline{\qquad}$

5. $58 \times 26 = \underline{\qquad}$

6. $56 \times 32 = \underline{\qquad}$

7. $87 \times 23 = \underline{\qquad}$

8. $95 \times 81 = \underline{\qquad}$

9. DIG DEEPER! Find 42×78 by breaking apart 42 first.

10. Modeling Real Life The Elephant Building is 335 feet high. A real Asian elephant is 12 feet tall. If 29 real elephants could stand on top of each other, would they reach the top of the building?

Elephant Building
Bangkok, Thailand

Review & Refresh

Find the difference. Then check your answer.

11. 30,698 − 5,439

12. 90,800 − 37,638

13. 214,507 − 73,569

Name _____

Use Partial Products to Multiply Two-Digit Numbers

4.5

Learning Target: Use place value and partial products to multiply.

Success Criteria:
- I can use place value to tell the value of each digit in a number.
- I can write the partial products for a multiplication problem.
- I can add the partial products to find a product.

Explore and Grow

How can you use the rectangles to find 24 × 53? Complete the equation.

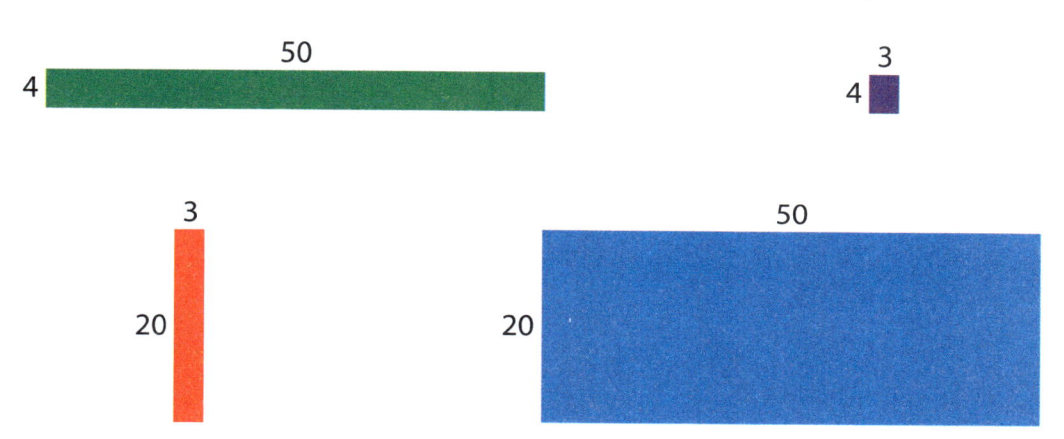

24 × 53 = _____

Reasoning What does the area of each rectangle represent?

Chapter 4 | Lesson 5

167

Think and Grow: Use Partial Products to Multiply Two-Digit Numbers

Example Use place value and partial products to find 27 × 48.

Estimate: 30 × 50 = _____

You can do these steps in any order.

Step 1: Multiply the tens by the tens.

```
   27
 × 48
 ────
  ☐   20 × 40
```

Step 2: Multiply the ones by the tens.

```
   27
 × 48
 ────
  800
  ☐   7 × 40
```

Step 3: Multiply the tens by the ones.

```
   27
 × 48
 ────
  800
  280
  ☐   20 × 8
```

Step 4: Multiply the ones by the ones.

```
   27
 × 48
 ────
  800
  280
  160
+ ☐   7 × 8
 ────
  ☐
```
Add the partial products.

So, 27 × 48 = _____.

Check: Because _____ is close to the estimate, _____, the answer is reasonable.

Show and Grow I can do it!

Find the product. Check whether your answer is reasonable.

1. Estimate: _____

```
     39
   × 15
   ────
     ☐
     ☐
     ☐
 +   ☐
```

2. Estimate: _____

```
     82
   × 63
   ────
     ☐
     ☐
     ☐
 +   ☐
```

3. Estimate: _____

```
     56
   × 71
   ────
     ☐
     ☐
     ☐
 +   ☐
```

Name _____

Multiply Two-Digit Numbers 4.6

Learning Target: Multiply two-digit numbers.
Success Criteria:
- I can multiply to find partial products.
- I can show how to regroup ones, tens, and hundreds.
- I can add partial products to find a product.

Explore and Grow

Use base ten blocks to model each product. Draw each model.

10 × 10 = _____	10 × 3 = _____
7 × 10 = _____	7 × 3 = _____

MP Reasoning How are the models related to the product 17 × 13?

Chapter 4 | Lesson 6   173

Think and Grow: Use Regrouping to Multiply Two-Digit Numbers

Example Find 87 × 64.

Estimate: 90 × 60 = _____

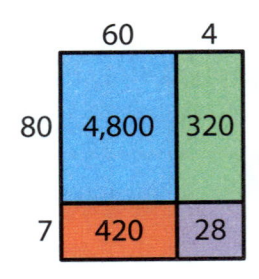

	60	4
80	4,800	320
7	420	28

Step 1: Multiply 87 by 4 ones, or 4.

```
      2
     87
   × 64
4 × 87 → 348
```

- 4 × 7 ones = 28 ones
  Regroup as 2 tens and 8 ones.
- 4 × 8 tens = 32 tens
  Add the regrouped tens: 32 + 2 = 34.

Step 2: Multiply 87 by 6 tens, or 60.

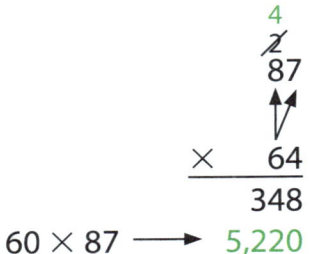

```
    4
    2
    87
  × 64
    348
60 × 87 → 5,220
```

- 6 tens × 7 = 42 tens
  Regroup as 4 hundreds and 2 tens or 20.
- 6 tens × 80 = 480 tens, or 48 hundreds
  Add the regrouped hundreds: 48 + 4 = 52.

So, 87 × 64 = _____.

Step 3: Add the partial products.

```
    4
    2
    87
  × 64
    348
 +5,220
  _____
```

Make sure the partial products are aligned in the correct place values.

Check: Because _____ is close to the estimate, _____, the answer is reasonable.

Show and Grow   I can do it!

Find the product. Check whether your answer is reasonable.

1. Estimate: _____

```
    41
  × 32
```

2. Estimate: _____

```
    □
    5 2
  × 4 6
```

3. Estimate: _____

```
    □
    □
    7 8
  × 3 5
```

174

Name _____

Homework & Practice 4.6

Learning Target: Multiply two-digit numbers.

Example Find 58 × 34.

Estimate: 60 × 35 = __2,100__

Think: 58 is 5 tens and 8 ones. 34 is 3 tens and 4 ones.

Step 1: Multiply 58 by 4 ones, or 4.

$$\begin{array}{r} \overset{3}{5}8 \\ \times\ 34 \\ \hline 232 \end{array}$$

4 × 58 → 232

Step 2: Multiply 58 by 3 tens, or 30.

$$\begin{array}{r} \overset{2}{\cancel{3}} \\ 58 \\ \times\ 34 \\ \hline 232 \\ 1{,}740 \end{array}$$

30 × 58 → 1,740

Step 3: Add the partial products.

$$\begin{array}{r} \overset{2}{\cancel{3}} \\ 58 \\ \times\ 34 \\ \hline 232 \\ +\ 1{,}740 \\ \hline \boxed{1{,}972} \end{array}$$

So, 58 × 34 = __1,972__.

Check: Because __1,972__ is close to the estimate, __2,100__, the answer is reasonable.

Find the product. Check whether your answer is reasonable.

1. Estimate: _____

$$\begin{array}{r} 31 \\ \times\ 92 \\ \hline \end{array}$$

2. Estimate: _____

$$\begin{array}{r} \square \\ 72 \\ \times\ 48 \\ \hline \end{array}$$

3. Estimate: _____

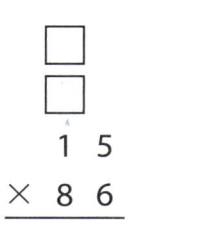

$$\begin{array}{r} \square \\ \square \\ 15 \\ \times\ 86 \\ \hline \end{array}$$

4. Estimate: _____

$$\begin{array}{r} 81 \\ \times\ 54 \\ \hline \end{array}$$

5. Estimate: _____

$$\begin{array}{r} 23 \\ \times\ 62 \\ \hline \end{array}$$

6. Estimate: _____

$$\begin{array}{r} 97 \\ \times\ 55 \\ \hline \end{array}$$

Chapter 4 | Lesson 6   177

Find the product. Check whether your answer is reasonable.

7. Estimate: _____

51 × 62 = _____

8. Estimate: _____

37 × 13 = _____

9. Estimate: _____

49 × 78 = _____

10. Newton plays 21 basketball games. He scores 12 points each game. How many points does he score in all?

11. **DIG DEEPER!** When you use regrouping to multiply two-digit numbers, why does the second partial product always end in 0?

12. **Number Sense** Find the missing digits.

```
        3 4
    × ☐ 5
    1 ☐ ☐
+ 2 , 0 4 0
  2 , 2 ☐ 0
```

13. **Modeling Real Life** A tiger dives 12 feet underwater. An otter dives 25 times deeper than the tiger. A walrus dives 262 feet underwater. Does the otter or walrus dive deeper?

Review & Refresh

14. Complete the table.

Standard Form	Word Form	Expanded Form
6,835		
		70,000 + 4,000 + 100 + 2
	five hundred one thousand, three hundred twenty-nine	

178

Name _____

Practice Multiplication Strategies 4.7

Learning Target: Use strategies to multiply two-digit numbers.

Success Criteria:
- I can choose a strategy to multiply.
- I can multiply two-digit numbers.
- I can explain the strategy I used to multiply.

Explore and Grow

Choose any strategy to find 60×80.

Multiplication Strategies
Place Value
Associative Property of Multiplication
Area Model
Distributive Property
Partial Products
Regrouping

Choose any strategy to find 72×13.

MP Reasoning Explain why you chose your strategies. Compare your strategies to your partner's strategies. How are they the same or different?

Chapter 4 | Lesson 7

Think and Grow: Practice Multiplication Strategies

Example Find 62 × 40.

One Way: Use place value.

62 × 40 = 62 × _____ tens

= _____ tens

= _____

So, 62 × 40 = _____.

Another Way: Use an area model and partial products.

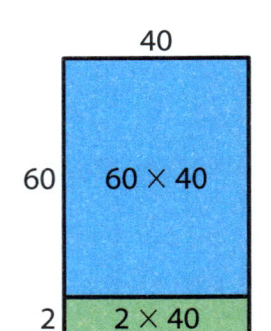

```
        40
   ┌─────────┐
60 │ 60 × 40 │   Add the partial products.
   │         │        ┌────┐
   │         │        │    │
   ├─────────┤       +├────┤
 2 │  2 × 40 │        │    │
   └─────────┘        └────┘
```

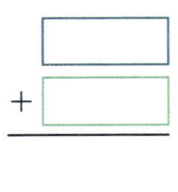

So, 62 × 40 = _____.

Example Find 56 × 83.

One Way: Use place value and partial products.

```
      56
    × 83
   ┌───┐
   │   │  80 × 50
   ├───┤
   │   │  80 × 6
   ├───┤
   │   │  3 × 50
   ├───┤
  +│   │  3 × 6
   ├───┤
   │   │
   └───┘
```

So, 56 × 83 = _____.

Another Way: Use regrouping.

Multiply 56 by 3 ones. Then multiply 56 by 8 tens. Regroup if necessary.

```
    □
    □
     5  6
  ×  8  3
```

So, 56 × 83 = _____.

Show and Grow   I can do it!

Find the product.

1. 90 × 37 = _____

2. 78 × 21 = _____

3. 14 × 49 = _____

Name _____

✓ Apply and Grow: Practice

Find the product.

4. 74 × 30 = _____

5. 51 × 86 = _____

6. 40 × 29 = _____

7. 92 × 80 = _____

8. 41 × 17 = _____

9. 60 × 53 = _____

Logic Find the missing factor.

10.
```
      72
  × [  ]
  ─────
      72
  + 2,880
  ─────
   2,952
```

11.
```
      65
  × [  ]
  ─────
     260
  + 1,950
  ─────
   2,210
```

12.
```
      93
  × [  ]
  ─────
     651
  + 7,440
  ─────
   8,091
```

13. **Writing** Explain why you start multiplying with the ones place when using regrouping to multiply.

14. **DIG DEEPER!** Find the missing digit so that both products are the same.

```
    26          3 0
  × 15        × 1 [ ]
```

Chapter 4 | Lesson 7   181

Think and Grow: Modeling Real Life

Example A swinging ship ride runs 50 times each afternoon. The ship has 10 rows of benches with 4 seats in each bench. If the ship is full each time it runs, how many people will ride the swinging ship in 1 afternoon?

Multiply to find how many people will ride the swinging ship ride each time.

$10 \times 4 =$ _____

_____ people will ride the swinging ship ride each time.

Multiply to find how many people will ride the swinging ship in 1 afternoon.

$40 \times 50 = 40 \times (5 \times 10)$ Rewrite 50 as 5×10.

$ = (40 \times 5) \times 10$ Associative Property of Multiplication

$ =$ _____ $\times 10$

$ =$ _____

_____ people will ride the swinging ship in 1 afternoon.

Show and Grow   I can think deeper!

15. A teacher orders 25 rock classification kits. Each kit has 4 rows with 9 rocks in each row. How many rocks are there in all?

16. A hotel has 12 floors with 34 rooms on each floor. 239 rooms are in use. How many rooms are *not* in use?

17. A child ticket for a natural history museum costs $13. An adult ticket costs twice as much as a child ticket. How much does it cost for 21 children and 14 adults to go to the museum?

Name _____

Homework & Practice 4.7

Learning Target: Use strategies to multiply two-digit numbers.

Example Find 17 × 80.

Use the Associative Property of Multiplication.

17 × 80 = 17 × (8 × 10)

= (17 × 8) × 10

= 136 × 10

= 1,360

So, 17 × 80 = __1,360__.

Example Find 34 × 52.

Use the Distributive Property.

34 × 52 = (30 + 4) × (50 + 2)

= (30 + 4) × 50 + (30 + 4) × 2

= (30 × 50) + (4 × 50) + (30 × 2) + (4 × 2)

= 1,500 + 200 + 60 + 8

= 1,768

So, 34 × 52 = __1,768__.

Find the product.

1. 16 × 13 = _____

2. 29 × 50 = _____

3. 78 × 45 = _____

4. 30 × 71 = _____

5. 62 × 14 = _____

6. 80 × 90 = _____

Chapter 4 | Lesson 7

183

Find the product.

7. 70 × 18 = _____

8. 32 × 59 = _____

9. 67 × 20 = _____

10. 51 × 84 = _____

11. 40 × 40 = _____

12. 23 × 97 = _____

13. Writing Which strategy do you prefer to use when multiplying two-digit numbers? Explain.

14. Patterns What number can you multiply the number of tires by to find the total weight? Use this pattern to complete the table.

Number of Tires	4	8	12	16	20
Total Weight (pounds)	80	160	240		

15. Modeling Real Life Each bag of popcorn makes 13 cups. A school has a movie day, and the principal brings 15 boxes of popcorn. Each box has 3 bags of popcorn. How many cups of popcorn does the principal bring?

Review & Refresh

Find the product.

16. 8 × 200 = _____

17. 7 × 300 = _____

18. 6,000 × 5 = _____

19. 9 × 90 = _____

20. 3,000 × 6 = _____

21. 5 × 500 = _____

Name _____

Problem Solving: Multiplication with Two-Digit Numbers

4.8

Learning Target: Solve multi-step word problems involving two-digit multiplication.

Success Criteria:
- I can understand a problem.
- I can make a plan to solve using letters to represent the unknown numbers.
- I can solve a problem using an equation.

Explore and Grow

Explain, in your own words, what the problem below is asking. Then explain how you can use multiplication to solve the problem.

A ferry can transport 64 cars each time it leaves a port. The ferry leaves a port 22 times in 1 day. How many cars can the ferry transport in 1 day?

MP **Construct Arguments** Make a plan to find how many cars the ferry can transport in 1 week.

Chapter 4 | Lesson 8   185

Think and Grow: Problem Solving: Multiplication with Two-Digit Numbers

Example A pet store receives a shipment of 8 boxes of dog treats. Each box is 2 feet high and has 18 bags of dog treats. How many ounces of dog treats does the pet store receive in the shipment?

Understand the Problem

What do you know?
- The store receives 8 boxes.
- Each box is 2 feet high.
- Each box has 18 bags of dog treats.
- Each bag weighs 32 ounces.

What do you need to find?
- You need to find how many ounces of dog treats the pet store receives in the shipment.

Make a Plan

How will you solve?
- Multiply 32 by 18 to find how many ounces of dog treats are in each box.
- Then multiply the product by 8 to find how many ounces of dog treats the pet store receives in the shipment.
- The height of each box is unnecessary information.

Solve

Step 1: How many ounces of dog treats are in each box?

$32 \times 18 = b$

b is the unknown product.

```
    ☐
   3 2
 × 1 8
 _____
```

$b = $ _____

Step 2: Use *b* to find how many ounces of dog treats the pet store receives.

$b \times 8 = p$

p is the unknown product.

```
  ☐
 ×   8
 _____
```

$p = $ _____

The pet store receives _____ ounces of dog treats.

Show and Grow I can do it!

1. Show how to solve the problem above using one equation.

186

Name _____

Homework & Practice 4.8

Learning Target: Solve multi-step word problems involving two-digit multiplication.

Example A store receives a shipment of 5 boxes of pretzels. Each box is 50 centimeters high and has 24 bags of pretzels. How many ounces of pretzels does the store receive in the shipment?

Think: What do you know? What do you need to find? How will you solve?

Step 1: How many ounces of pretzels are in each box?

$16 \times 24 = b$

b is the unknown product.

```
    1
    2
    1 6
 ×  2 4
   ─────
    6 4
 + 3 2 0
   ─────
   3 8 4
```

$b = \underline{384}$

Step 2: Use *b* to find how many ounces of pretzels the store receives.

$\underline{384} \times 5 = p$

p is the unknown product.

```
    4 2
    384
 ×    5
   ─────
  1,920
```

$p = \underline{1,920}$

The store receives __1,920__ ounces of pretzels.

Understand the problem. Then make a plan. How will you solve? Explain.

1. Seventy-two mushers compete in a sled-dog race. Each musher has 16 dogs. How many more dogs compete in the race than mushers?

2. A photographer buys 3 USB drives that each cost $5. She puts 16 folders on each drive. Each folder has 75 photographs. How many photographs does the photographer put on the USB drives in all?

Chapter 4 | Lesson 8

189

3. A teacher has 68 students take a 25-question test. The teacher checks the answers for 9 of the tests. How many answers does the teacher have left to check?

4. Each day, a cyclist bikes uphill for 17 miles and downhill for 18 miles. She drinks 32 fluid ounces of water after each bike ride. How many miles does the cyclist bike in 2 weeks?

5. **Precision** Which expressions can be used to solve the problem?

Twelve friends play a game that has 308 cards. Each player receives 16 cards. How many cards are left?

$(308 - 12) \times 16$ $308 - (16 \times 12)$

$308 - (12 \times 16)$ $(308 - 16) - 12$

6. **Modeling Real Life** A child ticket costs $14 less than an adult ticket. What is the total ticket cost for 18 adults and 37 children?

Adult Ticket: $57

7. **Modeling Real Life** An artist creates a pattern by alternating square and rectangular tiles. The pattern has 14 square tiles and 13 rectangular tiles. How long is the pattern?

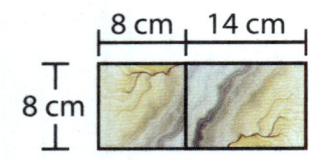

8 cm, 14 cm
8 cm

8. **Modeling Real Life** A cargo ship has 34 rows of crates. Each row has 16 stacks of crates. There are 5 crates in each stack. The ship workers unload 862 crates. How many crates are still on the ship?

Review & Refresh

Estimate the product.

9. 4×85

10. 6×705

11. $8 \times 7,923$

Name _____

Performance Task 4

Wind turbines convert wind to energy. Most wind turbines have 3 blades. The blades rotate slower or faster depending on the speed of the wind. More energy is generated when the blades spin faster.

1. A wind turbine rotates between 15 and 40 times in 1 minute.

   a. What is the least number of times the turbine rotates in 1 hour?

   b. What is the greatest number of times the turbine rotates in 1 hour?

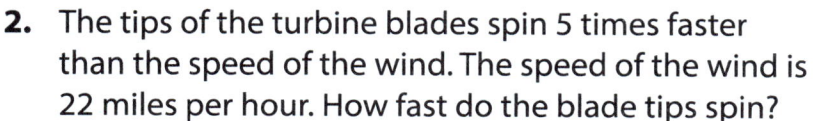

2. The tips of the turbine blades spin 5 times faster than the speed of the wind. The speed of the wind is 22 miles per hour. How fast do the blade tips spin?

3. A turbine farm has 7 large wind turbines. Each wind turbine can generate enough energy to power 1,485 houses. How many houses can the turbine farm power in all?

Wind Turbines	
Length of Each Blade (meters)	Number of Houses Powered
15	110
30	440

4. Use the chart.

   a. How many times more houses are powered when the length of each blade is doubled?

   b. A wind turbine has blades that are each 60 meters long. How many houses can the wind turbine power?

Chapter 4　　191

Multiplication Boss

Directions:

1. Each player flips 4 Number Cards and uses them in any order to create a multiplication problem with two-digit factors.
2. Each player finds the product of the two factors.
3. Players compare products. The player with the greater product takes all 8 cards.
4. If the products are equal, each player flips 4 more cards and plays again. The player with the greater product takes all 16 cards.
5. The player with the most cards at the end of the round wins!

Name _____

Chapter Practice 4

4.1 Multiply by Tens

Find the product.

1. 50 × 20 = _____
2. 30 × 60 = _____
3. 80 × 10 = _____

4. 40 × 70 = _____
5. 60 × 50 = _____
6. 90 × 90 = _____

7. 70 × 11 = _____
8. 18 × 30 = _____
9. 20 × 75 = _____

Find the missing factor.

10. 40 × _____ = 3,200
11. _____ × 20 = 1,200
12. 30 × _____ = 2,100

4.2 Estimate Products

Estimate the product.

13. 25 × 74
14. 16 × 28
15. 42 × 81

Open-Ended Write two possible factors that can be estimated as shown.

16. 8,100

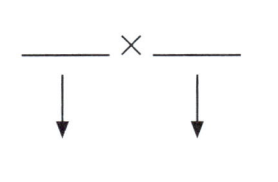

    _____ × _____
    ↓   ↓
    _____ × _____ = 8,100

17. 400

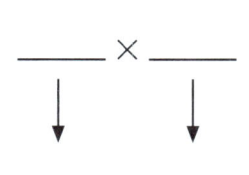

    _____ × _____
    ↓   ↓
    _____ × _____ = 400

Chapter 4 193

4.3 Use Area Models to Multiply Two-Digit Numbers

Draw an area model to find the product.

18. 13 × 19 = _____

19. 21 × 36 = _____

20. YOU BE THE TEACHER Your friend finds 28 × 24. Is your friend correct? Explain.

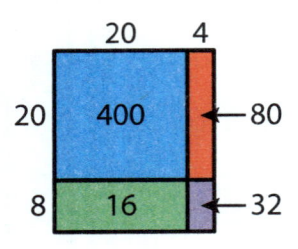

```
       20    4
    ┌──────┬──┐
 20 │ 400  │80│ ← 80
    ├──────┼──┤
  8 │  16  │32│ ← 32
    └──────┴──┘
```

400 + 16 + 80 + 32 = 528

4.4 Use Distributive Property to Multiply Two-Digit Numbers

Use the Distributive Property to find the product.

21. 27 × 34 = 27 × (30 + 4)

= (27 × 30) + (27 × 4)

= (20 + 7) × _____ + (20 + 7) × _____

= (20 × _____) + (7 × _____) + (20 × _____) + (7 × _____)

= _____ + _____ + _____ + _____

= _____

So, 27 × 34 = _____.

22. 43 × 18 = _____

23. 35 × 57 = _____

24. 81 × 76 = _____

4.5 Use Partial Products to Multiply Two-Digit Numbers

Find the product. Check whether your answer is reasonable.

25. 18
     × 22

26. 73
     × 46

27. 39
     × 84

28. 57 × 19 = _____

29. 38 × 65 = _____

30. 94 × 26 = _____

Reasoning Find the missing digits. Then find the product.

31.
```
      ☐ 2
   × ☐ 5
   ——————
     3 0 0
       6 0
       5 0
   +   1 0
   ——————
    ☐
```

32.
```
         ☐ 1
      × ☐ 4
      ——————
      2,0 0 0
          4 0
         2 0 0
      +      4
      ——————
       ☐
```

4.6 Multiply Two-Digit Numbers

Find the product. Check whether your answer is reasonable.

33. Estimate: _____

  62
× 34

34. Estimate: _____

  87
× 91

35. Estimate: _____

  73
× 45

36. Estimate: _____

13 × 21 = _____

37. Estimate: _____

42 × 53 = _____

38. Estimate: _____

29 × 66 = _____

Chapter 4   195

4.7 Practice Multiplication Strategies

Find the product.

39. 80 × 30 = _____

40. 26 × 51 = _____

41. 94 × 70 = _____

42. 15 × 67 = _____

43. 40 × 38 = _____

44. 29 × 92 = _____

45. Modeling Real Life A Ferris wheel runs 40 times each day. It has 16 cars with 4 seats in each car. If the Ferris wheel is full each time it runs, how many people will ride it in 1 day?

4.8 Problem Solving: Multiplication with Two-Digit Numbers

46. A music fan memorizes 59 songs for a concert. Her goal is to memorize all of the songs from 13 albums. There are 15 songs on each album. How many more songs does the music fan still need to memorize?

47. Find the area of the Jamaican flag.

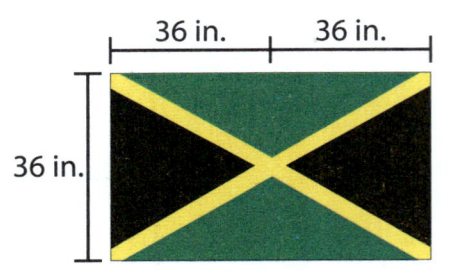

36 in. | 36 in.

36 in.

5 Divide Multi-Digit Numbers by One-Digit Numbers

- What is a planetarium?
- Your grade is going on a field trip to a planetarium. How can you use division to estimate the number of people in each tour group?

Chapter Learning Target:
Understand dividing one-digit numbers.

Chapter Success Criteria:
- I can divide a number.
- I can use division facts to estimate a quotient.
- I can write division problems.
- I can solve division problems.

5 Vocabulary

Name _____

Review Words
dividend
divisor
quotient

Organize It

Use the review words to complete the graphic organizer.

(_____) 10 ÷ 2 = **5** The answer when you divide one number by another number	(_____) **10** ÷ 2 = 5 The number of objects or the amount you want to divide
(_____) 10 ÷ **2** = 5 The number by which you divide	(Model) 10 ÷ 2 = 5

Division

Define It

Use your vocabulary cards to match.

1. partial quotients

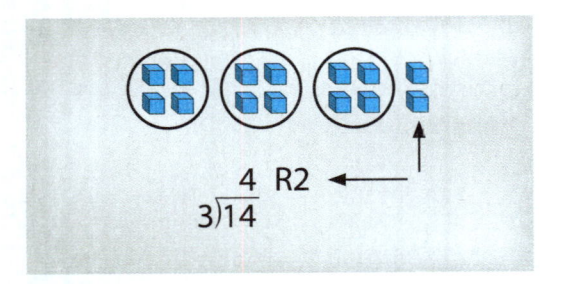

   4 R2
   3)14

2. remainder

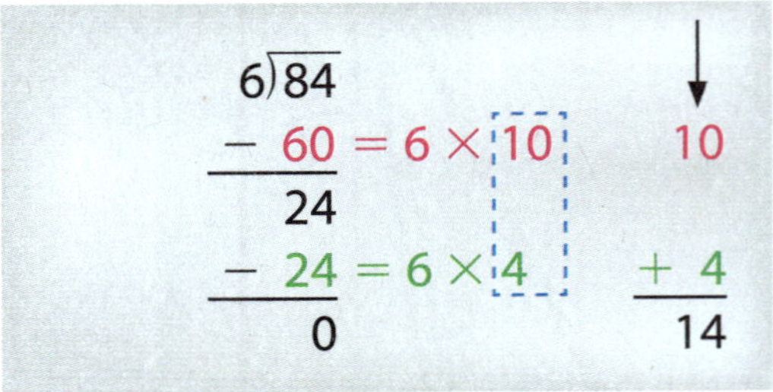

   6)84
   − 60 = 6 × 10   10
   24
   − 24 = 6 × 4   + 4
   0   14

198

Chapter 5 Vocabulary Cards

partial quotients

remainder

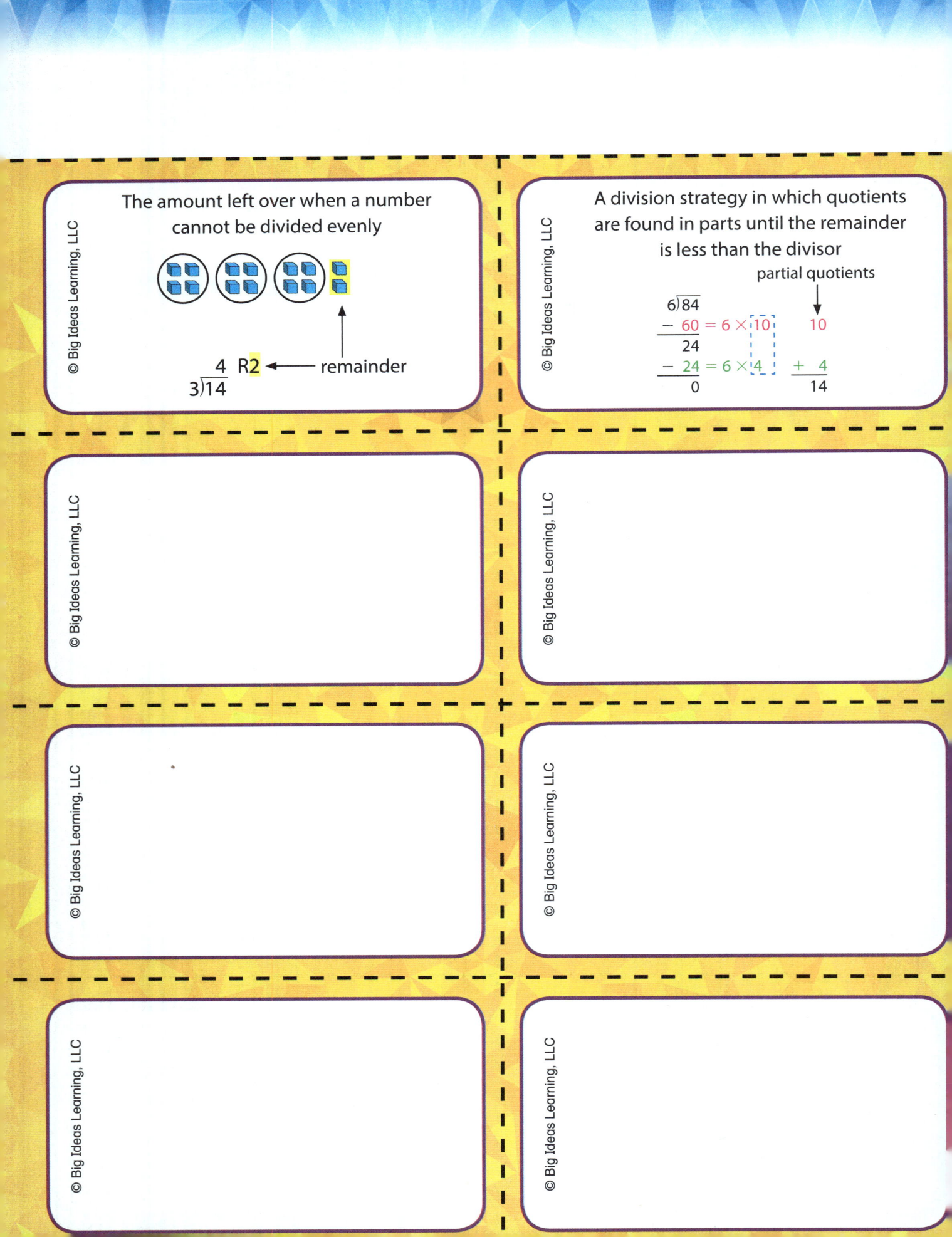

The amount left over when a number cannot be divided evenly

$$\begin{array}{r}4\ \text{R}2 \\ 3{\overline{\smash{\big)}\,14}}\end{array}$$ ← remainder

A division strategy in which quotients are found in parts until the remainder is less than the divisor

partial quotients

$$\begin{array}{r}6{\overline{\smash{\big)}\,84}} \\ -60 = 6 \times 10 \quad\quad 10 \\ \hline 24 \\ -24 = 6 \times 4 \quad\quad +\ 4 \\ \hline 0 \quad\quad\quad\quad\quad 14\end{array}$$

© Big Ideas Learning, LLC

Name _____

Divide Tens, Hundreds, and Thousands

5.1

Learning Target: Use place value to divide tens, hundreds, or thousands.

Success Criteria:
- I can divide a multiple of ten, one hundred, or one thousand by a one-digit number.
- I can explain how to use place value and division facts to divide tens, hundreds, or thousands.

Explore and Grow

Use a model to find each missing factor. Draw each model. Then write the related division equation.

____ × 2 = 8	____ × 2 = 80
____ × 2 = 800	____ × 2 = 8,000

What pattern do you notice?

MP Repeated Reasoning Explain how $12 \div 4$ can help you find $1,200 \div 4$.

Chapter 5 | Lesson 1

199

Think and Grow: Divide Tens, Hundreds, and Thousands

You can use place value and basic division facts to divide tens, hundreds, or thousands by one-digit numbers.

Example Find 270 ÷ 9.

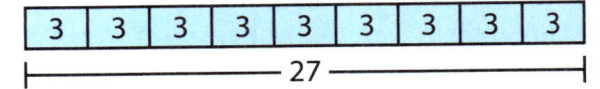

| 3 | 3 | 3 | 3 | 3 | 3 | 3 | 3 | 3 |

⊢——— 27 ———⊣

Think: 27 ÷ 9 = _____   Division fact

| 30 | 30 | 30 | 30 | 30 | 30 | 30 | 30 | 30 |

⊢——— 270 ———⊣

270 ÷ 9 = _____ tens ÷ 9   Use place value.

= _____ tens   Divide.

= _____

So, 270 ÷ 9 = _____.

Example Find 5,600 ÷ 8.

Think: 56 ÷ 8 = _____   Division fact

5,600 ÷ 8 = _____ hundreds ÷ 8   Use place value.

= _____ hundreds   Divide.

= _____

So, 5,600 ÷ 8 = _____.

Just as you use addition to check subtraction, use multiplication to check your answer.

Show and Grow   I can do it!

1. Find 2,400 ÷ 6.

   Think: _____ ÷ _____ = _____

   2,400 ÷ 6 = _____ hundreds ÷ 6

   = _____ hundreds

   = _____

   So, 2,400 ÷ 6 = _____.

2. Find each quotient.

   49 ÷ 7 = _____

   490 ÷ 7 = _____

   4,900 ÷ 7 = _____

Name _____

Homework & Practice 5.1

Learning Target: Use place value to divide tens, hundreds, or thousands.

Example Find 450 ÷ 5.

Think: 45 ÷ 5 = __9__

450 ÷ 5 = __45__ tens ÷ 5

 = __9__ tens

 = __90__

So, 450 ÷ 5 = __90__.

Example Find 2,800 ÷ 7.

Think: 28 ÷ 7 = __4__

2,800 ÷ 7 = __28__ hundreds ÷ 7

 = __4__ hundreds

 = __400__

So, 2,800 ÷ 7 = __400__.

1. Find 150 ÷ 3.

Think: _____ ÷ _____ = _____

150 ÷ 3 = _____ tens ÷ 3

 = _____ tens

 = _____

So, 150 ÷ 3 = _____.

2. Find 6,300 ÷ 7.

Think: _____ ÷ _____ = _____

6,300 ÷ 7 = _____ hundreds ÷ 7

 = _____ hundreds

 = _____

So, 6300 ÷ 7 = _____.

Find each quotient.

3. 12 ÷ 2 = _____

120 ÷ 2 = _____

1,200 ÷ 2 = _____

4. 40 ÷ 8 = _____

400 ÷ 8 = _____

4,000 ÷ 8 = _____

Find the quotient.

5. 80 ÷ 8 = _____

6. 300 ÷ 6 = _____

7. 1,000 ÷ 5 = _____

Chapter 5 | Lesson 1

203

Find the quotient.

8. 40 ÷ 4 = ____

9. 6,400 ÷ 8 = ____

10. 350 ÷ 5 = ____

11. 2,100 ÷ 7 = ____

12. 240 ÷ 3 = ____

13. 90 ÷ 9 = ____

DIG DEEPER! Find the missing number.

14. 50 ÷ ____ = 10

15. ____ ÷ 7 = 600

16. 320 ÷ ____ = 40

Compare.

17. 30 ÷ 3 ◯ 1 × 10

18. 560 ÷ 8 ◯ 7 × 100

19. 4,900 ÷ 7 ◯ 7 × 1,000

20. A movie theater has 180 seats. The seats are divided into 9 equal rows. How many seats are in each row?

21. **Number Sense** What is Newton's number?

> When I divide my number by 8, I get 60. When I divide my number by 6, I get 80.

22. **Modeling Real Life** A gorilla understands 2,000 words. She understands 4 times as many words as a toddler. How many words does the toddler understand?

Review & Refresh

Compare.

23. 1,834 ◯ 1,796

24. 62,905 ◯ 62,081

25. 9,142 ◯ 9,146

26. 52,048 ◯ 52,071

27. 402,157 ◯ 402,157

28. 387,402 ◯ 384,927

Name _____

Estimate Quotients 5.2

Learning Target: Use division facts and compatible numbers to estimate quotients.

Success Criteria:
- I can use division facts and compatible numbers to estimate a quotient.
- I can find two estimates that a quotient is between.

Explore and Grow

Explain how you can use the table to estimate 740 ÷ 8.

10 × 8 = _____	60 × 8 = _____
20 × 8 = _____	70 × 8 = _____
30 × 8 = _____	80 × 8 = _____
40 × 8 = _____	90 × 8 = _____
50 × 8 = _____	100 × 8 = _____

740 ÷ 8 is about _____.

MP **Reasoning** Why did you choose your estimate? Compare your results with your partner.

Chapter 5 | Lesson 2

205

Think and Grow: Estimate Quotients

You can use division facts and compatible numbers to estimate a quotient.

Example Estimate 154 ÷ 4.

Look at the first two digits of the dividend and use basic division facts.

> Compatible numbers are numbers that are easy to divide and are close to the actual numbers.

Think: What number close to 154 is easily divided by 4?

Try 120.   12 ÷ 4 = _____, so 120 ÷ 4 = _____.

Try 160.   16 ÷ 4 = _____, so 160 ÷ 4 = _____.

Choose 160 because 154 is closer to 160.

So, 154 ÷ 4 is about _____.

When solving division problems, you can check whether an answer is reasonable by finding two numbers that a quotient is between.

Example Find two numbers that the quotient 6,427 ÷ 7 is between.

> Look at the first two digits of the dividend and use basic division facts.

Think: What numbers close to 6,427 are easily divided by 7?

Use 6,300.   63 ÷ 7 = _____, so 6,300 ÷ 7 = _____.

Use 7,000.   70 ÷ 7 = _____, so 7,000 ÷ 7 = _____.

6,427 is between 6,300 and 7,000.

So, the quotient 6,427 ÷ 7 is between _____ and _____.

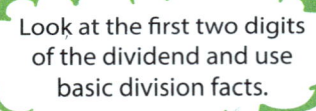

Show and Grow   I can do it!

Estimate the quotient.

1. 61 ÷ 3

2. 465 ÷ 9

Find two numbers that the quotient is between.

3. 477 ÷ 8

4. 5,194 ÷ 6

Name _____

Understand Division and Remainders

5.3

Learning Target: Use models to find quotients and remainders.

Success Criteria:
- I can use models to divide numbers that do not divide evenly.
- I can find a quotient and a remainder.
- I can interpret the quotient and the remainder in a division problem.

Explore and Grow

Use base ten blocks to determine whether 14 can be divided equally among 2, 3, 4, or 5 groups. Draw and describe your models.

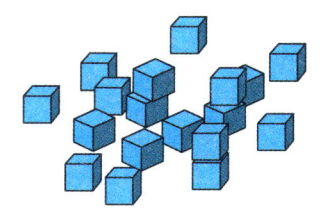

2 equal groups	3 equal groups
4 equal groups	5 equal groups

MP Structure Explain why the units that are left over cannot be put into a group.

Chapter 5 | Lesson 3

211

Think and Grow: Find and Interpret Remainders

Sometimes you cannot divide a number evenly and there is an amount left over.

The amount left over is called the **remainder**. Use an R to represent the remainder.

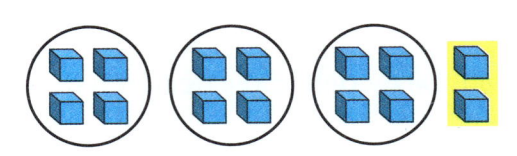

$14 \div 3 = 4$ with 2 left over
$14 \div 3 = 4$ R2

$$3\overline{)14} \quad 4 \text{ R2}$$

Example Find $27 \div 4$.

Divide 27 into 4 equal groups.

You need to regroup 2 tens as 20 ones.

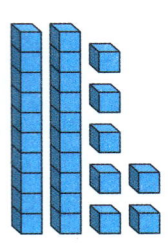

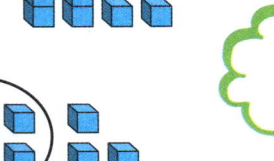

Think: $27 \div 4 = 6$ R3 because $4 \times 6 + 3 = 27$.

Number of units in each group: _____

Number of units left over: _____

So, $27 \div 4 =$ _____ R _____.

Show and Grow   I can do it!

Use a model to find the quotient and the remainder.

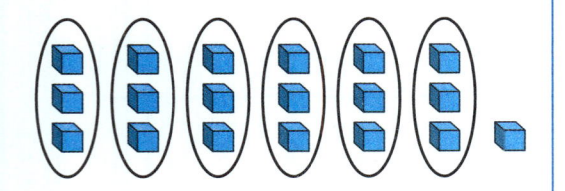

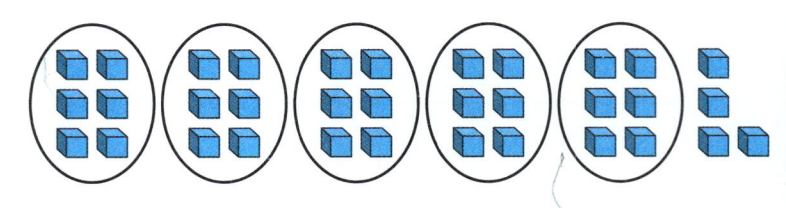

1. $19 \div 6 =$ _____ R _____

2. $34 \div 5 =$ _____ R _____

3. $26 \div 3 =$ _____ R _____

4. $20 \div 7 =$ _____ R _____

Name _____

Homework & Practice 5.3

Learning Target: Use models to find quotients and remainders.

Example Find 20 ÷ 3.

Divide 20 into 3 equal groups.

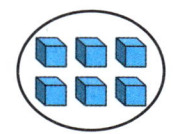

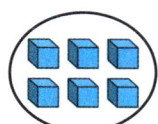

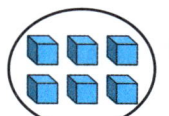

Number of units in each group: __6__

Number of units left over: __2__

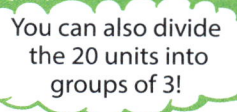

So, 20 ÷ 3 = __6__ R __2__.

You can also divide the 20 units into groups of 3!

Use a model to find the quotient and the remainder.

1. 25 ÷ 7 = _____ R _____

2. 19 ÷ 2 = _____ R _____

3. 27 ÷ 6 = _____ R _____

4. 26 ÷ 4 = _____ R _____

Chapter 5 | Lesson 3

Use a model to find the quotient and the remainder.

5. 29 ÷ 8 = _____ R _____

6. 11 ÷ 2 = _____ R _____

7. **DIG DEEPER!** A number divided by 4 has a remainder. What numbers might the remainder be? Explain.

8. **Modeling Real Life** Tours of a space center can have no more than 7 guests. There are 31 guests in line to tour the space center.
- How many tours are full?
- How many tours are needed?
- How many guests are in the last tour?

9. **Modeling Real Life** You need 3 googly eyes to make one monster puppet. You have 28 googly eyes. How many monster puppets can you make?

10. **Modeling Real Life** Forty-one students attend tryouts for a debate league. Each team can have 6 students. How many students will *not* be on a team?

11. **Modeling Real Life** A book has 37 pages. You read 7 pages each day. How many days will it take you to finish the book?

Review & Refresh

Estimate the sum or difference.

12. 50,917 − 23,846 = _____

13. 499,042 + 181,765 = _____

Name _____

Use Partial Quotients 5.4

Learning Target: Use partial quotients to divide.
Success Criteria:
- I can explain how to use an area model to divide.
- I can write partial quotients for a division problem.
- I can add the partial quotients to find a quotient.

Explore and Grow

Use the area models to find 3 × 12 and 36 ÷ 3.

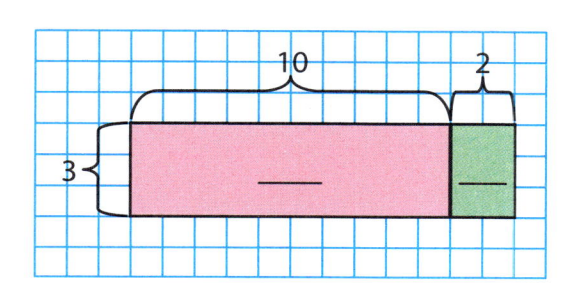

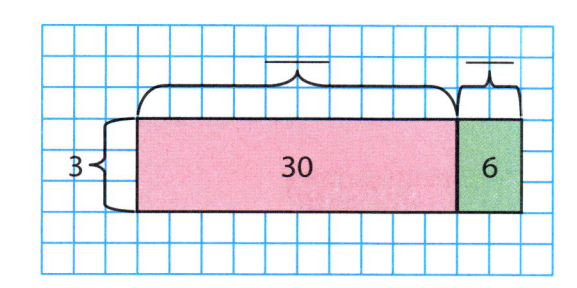

[Area model: 3 by (10 + 2), pink section labeled 10 wide, green section labeled 2 wide]

3 × 12 = _____

[Area model: 3 by (? + ?), pink section = 30, green section = 6]

36 ÷ 3 = _____

Reasoning How does the Distributive Property relate to each of the area models? Explain.

Chapter 5 | Lesson 4   217

Think and Grow: Use Partial Quotients to Divide

To divide using **partial quotients**, subtract a multiple of the divisor that is less than the dividend. Continue to subtract multiples until the remainder is less than the divisor. The factors that are multiplied by the divisor are called partial quotients. Their sum is the quotient.

Example Use an area model and partial quotients to find $235 \div 5$.

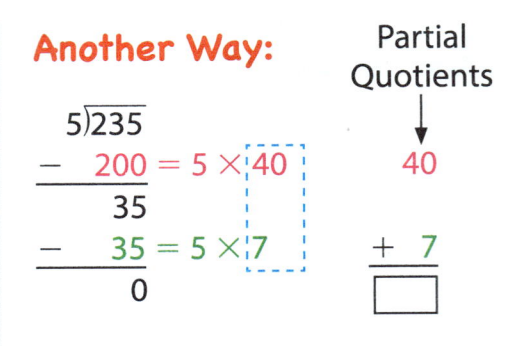

One Way:

$$\begin{array}{r} 5\overline{)235} \\ -\ 100 = 5 \times 20 \\ \hline 135 \\ -\ 100 = 5 \times 20 \\ \hline 35 \\ -\ 35 = 5 \times 7 \\ \hline 0 \end{array}$$

Partial Quotients:
20
20
+ 7
____

So, $235 \div 5 =$ _____.

5	20	20	7
	100	100	35

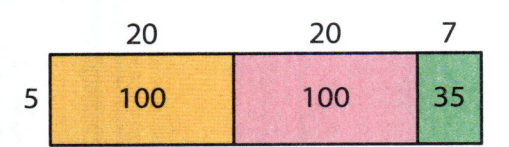

Area = 235 square units

Another Way:

$$\begin{array}{r} 5\overline{)235} \\ -\ 200 = 5 \times 40 \\ \hline 35 \\ -\ 35 = 5 \times 7 \\ \hline 0 \end{array}$$

Partial Quotients:
40
+ 7
____

So, $235 \div 5 =$ _____.

5	40	7
	200	35

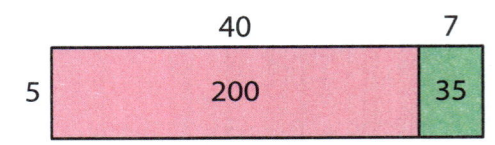

Area = 235 square units

Show and Grow   I can do it!

Use an area model and partial quotients to divide.

1. $60 \div 4 =$ _____

$$\begin{array}{r} 4\overline{)60} \\ -\ 40 = 4 \times \underline{} \\ \hline \Box \\ -\ 20 = 4 \times \underline{} \\ \hline \Box \end{array}$$

$\Box$
$+\ \Box$
____

4	40	20

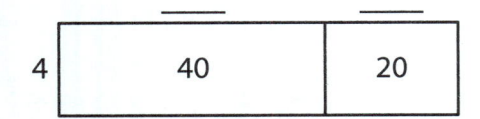

2. $192 \div 3 =$ _____

$$\begin{array}{r} 3\overline{)192} \\ -\ \Box = 3 \times \underline{} \\ \hline \Box \\ -\ \Box = 3 \times \underline{} \\ \hline \Box \end{array}$$

$\Box$
$+\ \Box$
____

3	180	___

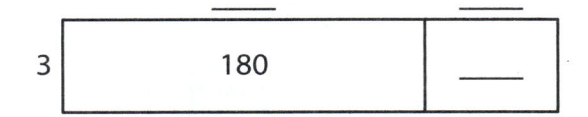

Name _____

Use Partial Quotients with a Remainder
5.5

Learning Target: Use partial quotients to divide and find remainders.

Success Criteria:
- I can use partial quotients to divide.
- I can find a remainder.

Explore and Grow

Use an area model to find 125 ÷ 5.

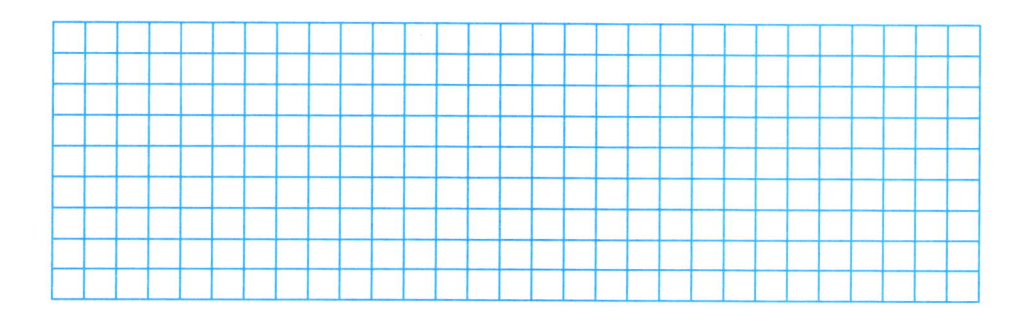

Can you use an area model to find 128 ÷ 5? Explain your reasoning.

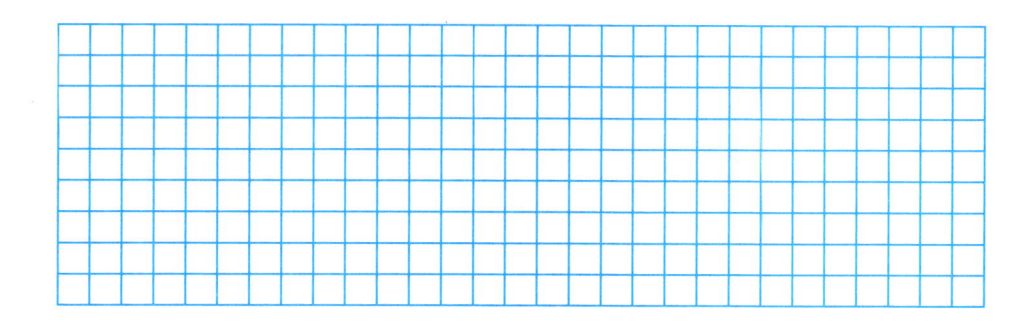

MP Construct Arguments Explain to your partner how your model shows that 5 does *not* divide evenly into 128.

Chapter 5 | Lesson 5

223

Think and Grow: Use Partial Quotients with a Remainder

Example Use partial quotients to find 2,918 ÷ 4.

> Continue to divide until the remainder is less than the divisor.

```
  4) 2,918
   − 2,800  = 4 × 700
       118
   −   100  = 4 × 25
        18
   −    16  = 4 × 4
         2
```

☐
☐
+ ☐
─────
☐ R ☐

So, 2,918 ÷ 4 = _____ R _____.

Show and Grow I can do it!

Use partial quotients to divide.

1. 82 ÷ 3 = _____

```
  3) 82
   − 60 = 3 × ___
     ☐
   − 21 = 3 × ___
     ☐
```

☐
+ ☐
───
☐ R ☐

2. 754 ÷ 9 = _____

```
  9) 754
   − 720 = 9 × ___
     ☐
   −  27 = 9 × ___
     ☐
```

☐
+ ☐
───
☐ R ☐

3. 8) 460

4. 5) 3,242

5. 6) 5,850

224

Name _____

Divide Two-Digit Numbers by One-Digit Numbers
5.6

Learning Target: Divide two-digit numbers by one-digit numbers.

Success Criteria:
- I can divide to find the partial quotients.
- I can show how to regroup 1 or more tens.
- I can use place value to record the partial quotients.

Explore and Grow

Use a model to find each quotient. Draw each model.

$$84 \div 4$$

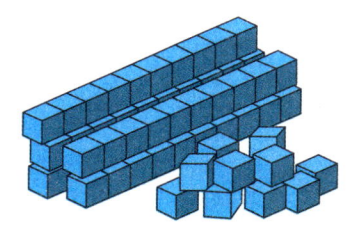

$$85 \div 5$$

MP **Construct Arguments** Explain to your partner how your methods for finding the quotients above are the same. Then explain how they are different.

Chapter 5 | Lesson 6

229

Think and Grow: Use Regrouping to Divide

Example Find $79 \div 3$.   Think: 79 is 7 tens and 9 ones.

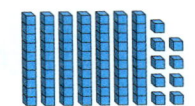

Divide the tens.

$$\begin{array}{r} 2 \\ 3\overline{)79} \\ -6 \\ \hline 1 \end{array}$$

Divide: 7 tens $\div$ 3
Multiply: 2 tens $\times$ 3
Subtract: 7 tens $-$ 6 tens
There is 1 ten left over.

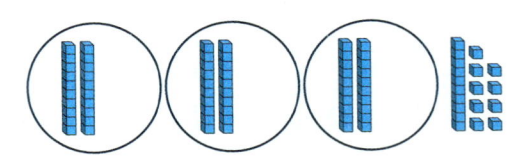

Regroup.

$$\begin{array}{r} 2 \\ 3\overline{)79} \\ -6\downarrow \\ \hline 19 \end{array}$$

Regroup 1 ten as 10 ones.
10 ones + 9 ones = 19 ones

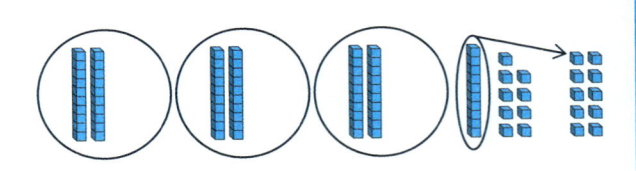

Divide the ones.

$$\begin{array}{r} 26 \text{ R1} \\ 3\overline{)79} \\ -6\downarrow \\ \hline 19 \\ -18 \\ \hline 1 \end{array}$$

Divide: 19 ones $\div$ 3
Multiply: 6 ones $\times$ 3
Subtract: 19 ones $-$ 18 ones
There is 1 one left over.

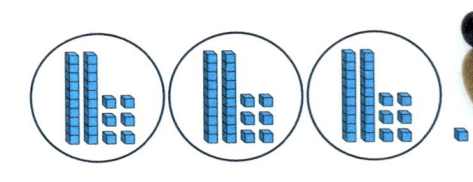

Check:
$26 \times 3 + 1 = 78 + 1 = 79$

So, $79 \div 3 =$ _____ R _____.

Show and Grow   I can do it!

Divide. Then check your answer.

1. $6\overline{)96}$

2. $2\overline{)88}$

3. $5\overline{)74}$ R _____

230

Name _____

Homework & Practice 5.6

Learning Target: Divide two-digit numbers by one-digit numbers.

Example Find 95 ÷ 4.   Think: 95 is 9 tens and 5 ones.

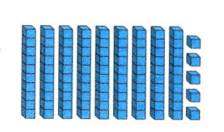

Divide the tens.

```
   2
4)95
 − 8
   1
```

Divide: 9 tens ÷ 4
Multiply: 2 tens × 4
Subtract 9 tens − 8 tens
There is 1 ten left over.

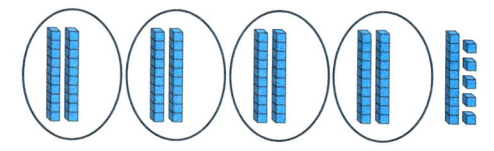

Regroup.

```
   2
4)95
 − 8↓
   15
```

Regroup 1 ten as 10 ones.
10 ones + 5 ones = 15 ones

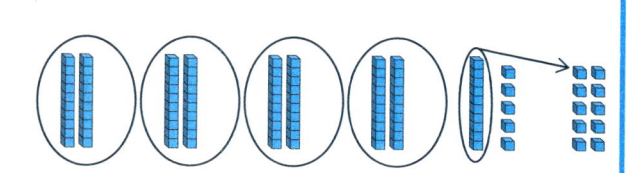

Divide the ones.

```
   23  R3
4)95
 − 8↓
   15
 − 12
    3
```

Divide: 15 ones ÷ 4
Multiply: 3 ones × 4
Subtract: 15 ones − 12 ones
There are 3 ones left over.

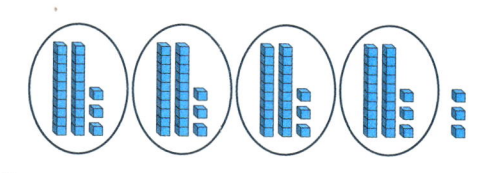

So, 95 ÷ 4 = __23__ R __3__

Divide. Then check your answer.

1.
```
    □
5)85
 − □↓
   □
 − □
   □
```

2.
```
   □□
3)63
```

3.
```
   □□  R ____
7)94
```

Chapter 5 | Lesson 6   233

Divide. Then check your answer.

4. 6)74

5. 8)92

6. 3)50

7. 2)83

8. 9)72

9. 7)65

10. Which One Doesn't Belong? Which problem does *not* require regrouping to solve?

2)36   3)55   2)47   4)92

11. Modeling Real Life A team of 6 students finishes an obstacle course in 66 minutes. Each student spends the same number of minutes on the course. How many minutes is each student on the course?

12. DIG DEEPER! You want to make 40 origami animals in 3 days. You want to make about the same number of animals each day. How many animals should you make each day? How can you interpret the remainder?

Review & Refresh

Estimate the product.

13. 32×67

14. 24×51

15. 96×75

Name _____

Divide Multi-Digit Numbers by One-Digit Numbers

5.7

Learning Target: Divide multi-digit numbers by one-digit numbers.

Success Criteria:
- I can use place value to divide.
- I can show how to regroup thousands, hundreds, or tens.
- I can find a quotient and a remainder.

Explore and Grow

Use a model to divide.
Draw each model.

$$348 \div 3$$

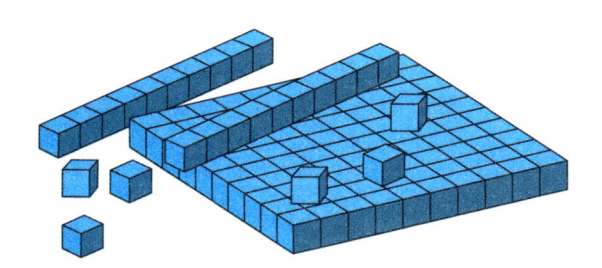

$$148 \div 3$$

MP Reasoning Explain why the quotient of $148 \div 3$ does *not* have a digit in the hundreds place.

Chapter 5 | Lesson 7

Think and Grow: Use Regrouping to Divide

Example Find 907 ÷ 5.

Estimate: 1,000 ÷ 5 = _____

Divide the hundreds.

```
   1
5)907
 - 5
   4
```

9 hundreds ÷ 5
1 hundred × 5
9 hundreds − 5 hundreds
There are 4 hundreds left over.

Divide the tens.

```
  18
5)907
 - 5↓
   40
 - 40
    0
```

Regroup 4 hundreds as 40 tens: 40 ÷ 5
8 tens × 5
40 tens − 40 tens
There are 0 tens left over.

Divide the ones.

```
  181 R2
5)907
 - 5
   40
 - 40↓
    07
   - 5
     2
```

There are no tens to regroup, so divide the ones: 7 ones ÷ 5
1 one × 5
7 ones − 5 ones
There are 2 ones left over.

So, 907 ÷ 5 = _____ R _____.

Check: Because _____ R _____ is close to the estimate, the answer is reasonable.

Show and Grow   I can do it!

Divide. Then check your answer.

1. ☐ R _____
    4)531

2. ☐
    5)7,180

3. ☐ R _____
    7)8,385

236

Name _____

Apply and Grow: Practice

Divide. Then check your answer.

4. 5)6,381

5. 3)4,605

6. 6)820

7. 6)7,039

8. 4)855

9. 2)367

10. 8)9,692

11. 7)8,345

12. 7)971

13. There are 8,274 people at an air show. The people are divided equally into 6 sections. How many people are in each section?

14. **YOU BE THE TEACHER** Newton finds 120 ÷ 5. Is he correct? Explain.

```
       114
   5)120
    − 5
      7
    − 5
     20
   − 20
      0
```

Chapter 5 | Lesson 7
237

Think and Grow: Modeling Real Life

Example There are 1,014 toy car tires at a factory. Each car needs 4 tires. How many toy cars can the factory workers make with the tires?

Each car needs 4 tires, so find 1,014 ÷ 4.

1 thousand *cannot* be shared among 4 groups without regrouping. So, regroup 1 thousand as 10 hundreds.

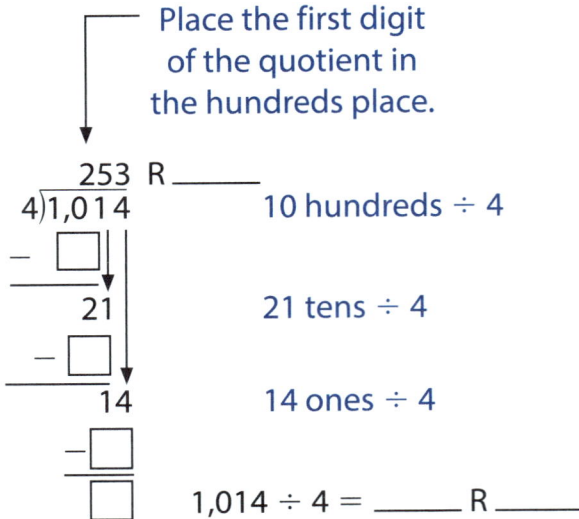

Place the first digit of the quotient in the hundreds place.

```
    253  R ____
4)1,014
  -☐
    21            10 hundreds ÷ 4
   -☐
     14           21 tens ÷ 4
    -☐
     ☐            14 ones ÷ 4
```

1,014 ÷ 4 = ____ R ____

Interpret the quotient and the remainder.

The quotient is ____. The factory workers can make ____ toy cars.

The remainder is ____. There are ____ tires left over.

Show and Grow   I can think deeper!

15. A principal orders 750 tablets. The distributor can fit 8 tablets in each box. How many boxes are needed to ship all of the tablets?

16. An athlete's heart rate after a 5-mile run is 171 beats per minute, which is 3 times as fast as her resting heart rate. What is the athlete's resting heart rate?

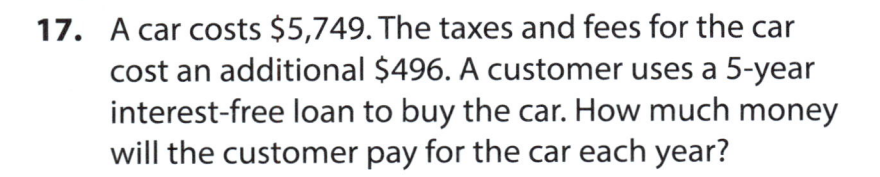

17. A car costs $5,749. The taxes and fees for the car cost an additional $496. A customer uses a 5-year interest-free loan to buy the car. How much money will the customer pay for the car each year?

Name _____

Divide by One-Digit Numbers
5.8

Learning Target: Divide by one-digit numbers.
Success Criteria:
- I can use place value to divide.
- I can explain why there might be a 0 in the quotient.
- I can find a quotient and a remainder.

Explore and Grow

Use a model to find each quotient. Draw each model.

$$312 \div 3$$

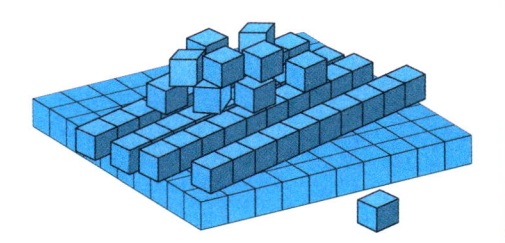

$$312 \div 4$$

MP

Structure Compare your models for each quotient. What is the same? What is different? What do you think this means when using regrouping to divide?

Chapter 5 | Lesson 8

241

Think and Grow: Divide by One-Digit Numbers

Example Find 4,829 ÷ 8.

4 thousands *cannot* be shared among 8 groups without regrouping. So, regroup 4 thousands as 40 hundreds and combine with 8 hundreds.

Divide the hundreds.

$$\begin{array}{r} 6 \\ 8{\overline{\smash{\big)}\,4{,}829}} \\ -\,48 \\ \hline 0 \end{array}$$

48 hundreds ÷ 8
6 hundreds × 8
48 hundreds − 48 hundreds
There are 0 hundreds left over.

Divide the tens.

$$\begin{array}{r} 60 \\ 8{\overline{\smash{\big)}\,4{,}829}} \\ -\,48\downarrow \\ \hline 02 \\ -0 \\ \hline 2 \end{array}$$

2 tens *cannot* be shared among 8 groups without regrouping. So, place a zero in the quotient.

0 tens × 8
2 tens − 0 tens
There are 2 tens left over.

> When a place value in the dividend cannot be divided by the divisor without regrouping, write a zero in the quotient.

Divide the ones.

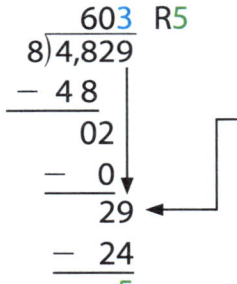

$$\begin{array}{r} 603\ \text{R}5 \\ 8{\overline{\smash{\big)}\,4{,}829}} \\ -\,48 \\ \hline 02 \\ -0\downarrow \\ \hline 29 \\ -\,24 \\ \hline 5 \end{array}$$

Regroup 2 tens as 20 ones and combine with 9 ones.

29 ones ÷ 8
3 ones × 8
29 ones − 24 ones
There are 5 ones left over.

So, 4,829 ÷ 8 = _____ R _____ .

Show and Grow I can do it!

Divide. Then check your answer.

1. 7)756

2. 6)364   R _____

3. 3)3,190   R _____

242

Name _____

Apply and Grow: Practice

Divide. Then check your answer.

4.
2)81

5.
4)428

6.
6)842

7.
3)2,724

8.
9)635

9.
6)1,442

10.
6)303

11.
5)2,530

12.
8)7,209

13. The 5 developers of a phone app earn a profit of $4,535 this month. They divide the profit equally. How much money does each developer get?

14. **YOU BE THE TEACHER** Newton finds 817 ÷ 4. Is he correct? Explain.

```
    24   R1
4)817
  − 8
    17
  − 16
     1
```

Chapter 5 | Lesson 8

243

Think and Grow: Modeling Real Life

Example Seven players are placed on each basketball team. Remaining basketball players are added to the teams, so some of the teams have 8 players. How many basketball teams have 7 players? 8 players?

Sport	Number of Players
Soccer	2,476
Basketball	1,839
Flag football	3,214
Ball hockey	952

There are 1,839 players signed up for basketball, so find 1,839 ÷ 7.

1 thousand *cannot* be shared among 7 groups without regrouping. So, regroup 1 thousand as 10 hundreds and combine with 8 hundreds.

```
    ☐  R ____
7)1,839           18 hundreds ÷ 7
 - ☐
    43            43 tens ÷ 7
  - ☐
     19           19 ones ÷ 7
    - ☐
     ☐            1,839 ÷ 7 = ____ R ____
```

Interpret the quotient and the remainder.

The quotient is ____. So, there are ____ basketball teams in all.

The remainder is ____. So, ____ basketball teams have 8 players.

Subtract to find how many teams have 7 players.   ____ − ____ = ____

So, ____ basketball teams have 7 players and ____ have 8 players.

Show and Grow   I can think deeper!

Use the table above.

15. Nine players are placed on each ball hockey team. Remaining players are added to the teams, so some of the teams have 10 players. How many ball hockey teams have 9 players? 10 players?

16. Eighty-four players who signed up to play soccer decide not to play. Eight players are placed on each soccer team. How many soccer teams are there?

Name _____

Homework & Practice 5.8

Learning Target: Divide by one-digit numbers.

Example Find $283 \div 4$.

2 hundreds *cannot* be shared among 4 groups without regrouping. So, regroup 2 hundreds as 20 tens and combine with 8 tens.

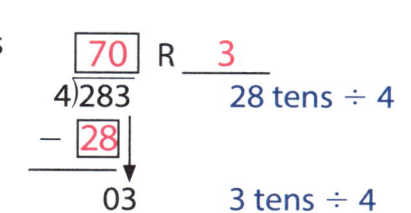

```
      70    R  3
   ┌─────          28 tens ÷ 4
  4)283
   − 28
      ──
       03         3 tens ÷ 4
     −  0
        ──
         3
```

When a place value in the dividend cannot be divided by the divisor without regrouping, write a zero in the quotient.

So, $283 \div 4 =$ __70__ R __3__ .

Divide. Then check your answer.

1. 8)832

2. 7)215   R ____

3. 5)5,078   R ____

4. 7)94

5. 6)731

6. 4)6,514

7. 3)62

8. 5)548

9. 2)4,136

Chapter 5 | Lesson 8   245

Divide. Then check your answer.

10. 7)214

11. 4)321

12. 6)5,162

13. 2)7,301

14. 5)603

15. 3)6,082

16. There are 450 pounds of grapes for a grape stomping contest. They are divided equally into 5 barrels. How many pounds of grapes are in each barrel?

17. DIG DEEPER! How could you change the dividend in Exercise 11 so that there would be no remainder? Explain.

18. Modeling Real Life Five actresses are placed on each team. Remaining actresses are added to the teams, so some of the teams have 6 actresses. How many teams have 5 actresses? 6 actresses?

Celebrity Game Show Sign-Ups	
Celebrity	Number of Players
Singers	105
Musicians	81
Actors	197
Actresses	202

Review & Refresh

Write the value of the underlined digit.

19. 8<u>6</u>,109

20. <u>1</u>5,327

21. <u>9</u>14,263

22. 284,5<u>0</u>5

Name _____

Problem Solving: Division 5.9

Learning Target: Solve multi-step word problems involving division.

Success Criteria:
- I can understand a problem.
- I can make a plan to solve using letters to represent the unknown numbers.
- I can solve a problem using an equation.

Explore and Grow

Make a plan to solve the problem.

A fruit vendor has 352 green apples and 424 red apples. The vendor uses all of the apples to make fruit baskets. He puts 8 apples in each basket. How many fruit baskets does the vendor make?

MP **Make Sense of Problems** The vendor decides that each basket should have 8 of the same colored apples. Does this change your plan to solve the problem? Will this change the answer? Explain.

Chapter 5 | Lesson 9

Think and Grow: Problem Solving: Division

Example The speed of sound in water is 1,484 meters per second. Sound travels 112 more than 4 times as many meters per second in water as it does in air. What is the speed of sound in air?

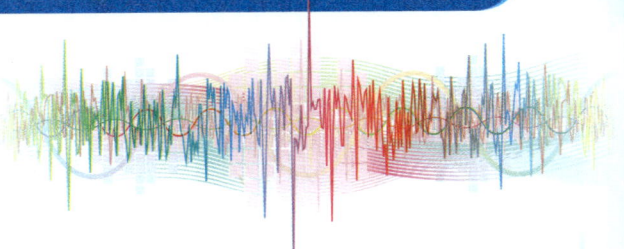

Understand the Problem

What do you know?
- The speed of sound in water is 1,484 meters per second.
- Sound travels 112 more than 4 times as many meters per second in water as it does in air.

What do you need to find?
- You need to find the speed of sound in air.

Make a Plan

How will you solve?
- Subtract 112 from 1,484 to find 4 times the speed of sound in air.
- Then divide the difference by 4 to find the speed of sound in air.

Solve

Step 1: $1,484 - 112 = d$

d is the unknown difference.

$$\begin{array}{r} 1,484 \\ -112 \\ \hline \square \end{array}$$

$d =$ _____

Step 2: $d \div 4 = a$

a is the unknown value.

$4\overline{)\square}$ with $\square$ on top

$a =$ _____

The speed of sound in air is _____ meters per second.

Show and Grow *I can do it!*

1. Explain how you can check whether your answer above is reasonable.

Name _____

✓ Apply and Grow: Practice

Understand the problem. What do you know? What do you need to find? Explain.

2. A surf shop owner divides 635 stickers evenly among all of her surfboards. Each surfboard has 3 tiki stickers and 2 turtle stickers. How many surfboards does she have?

3. There are 1,008 projects in a science fair. The projects are divided equally into 9 rooms. Each room has 8 equal rows of projects. How many projects are in each row?

Understand the problem. Then make a plan. How will you solve? Explain.

4. Of 78 students who work on a mural, 22 students design it, and the rest of the students paint it. The painters are divided equally among 4 areas of the mural. How many painters are assigned to each area?

5. The Winter Olympics occur twice every 8 years. How many times will the Winter Olympics occur in 200 years?

6. A party planner wants to put 12 balloons at each of 15 tables. The balloons come in packages of 8. How many packages of balloons must the party planner buy?

7. An art teacher has 8 boxes of craft sticks. Each box has 235 sticks. The students use the sticks to make as many hexagons as possible. How many sticks are *not* used?

Chapter 5 | Lesson 9

249

Think and Grow: Modeling Real Life

Example A book enthusiast has $200 to buy an e-reader and e-books. He uses a $20 off coupon and buys the e-reader shown. Each e-book costs $6. How many e-books can the book enthusiast buy?

Think: What do you know? What do you need to find? How will you solve?

$119

Step 1: How much money does the book enthusiast pay for the e-reader?

Subtract $20 from $119.

```
  119
−  20
─────
```

Step 2: Subtract to find how much money he has left to spend on e-books.

$200 − _____ = d

d is the unknown difference.

```
  200
−  ☐
────
   ☐
```

d = _____

Step 3: Use *d* to find the number of e-books the book enthusiast can buy.

$d \div 6 = q$

q is the unknown value.

☐ R _____
6)☐

q = _____ R _____

So, the book enthusiast can buy _____ e-books.

Show and Grow I can think deeper!

8. You run 17 laps around a track. Newton runs 5 times as many laps as you. Descartes runs 35 more laps than Newton. Eight laps around the track are equal to 1 mile. How many miles does Descartes run?

Name _____

Homework & Practice 5.9

Learning Target: Solve multi-step word problems involving division.

Example The Statue of Liberty was a gift from the people of France to the people of the United States of America. The Eiffel Tower in France is 986 feet tall. It is 234 feet shorter than 4 times the height of the Statue of Liberty. How tall is the Statue of Liberty?

Eiffel Tower   Statue of Liberty

Think: What do you know? What do you need to find? How will you solve?

Step 1: $986 + 234 = x$

x is the unknown sum.

```
  986
+ 234
─────
1,220
```
$x = $ __1,220__

Step 2: $x \div 4 = y$

y is the unknown quotient.

$4\overline{)1{,}220}$ = __305__

$y = $ __305__

The Statue of Liberty is __305__ feet tall.

Understand the problem. Then make a plan. How will you solve? Explain.

1. You borrow a 235-page book from the library. You read 190 pages. You have 3 days left until you have to return the book. You want to read the same number of pages each day to finish the book. How many pages should you read each day?

2. There are 24 fourth graders and 38 fifth graders traveling to a math competition. If 8 students can fit into each van, how many vans are needed?

Chapter 5 | Lesson 9

3. Your class has 3 bags of buttons to make riding horses for a relay race. Each bag has 54 buttons. What is the greatest number of horses your class can make?

Each horse needs 4 buttons.

4. Factory workers make 2,597 small, 2,597 medium, and 2,597 large plush toys. The workers pack the toys into boxes with 4 toys in each box. How many toys are left over?

5. **Writing** Write and solve a two-step word problem that can be solved using division.

6. **Modeling Real Life** You exercise for 300 minutes this week. Outside of jogging, you divide your exercising time equally among 3 other activities. How many minutes do you spend on each of your other 3 activities?

M: jog 35 min
W: jog 35 min
F: jog 35 min

7. **Modeling Real Life** Drones are used to help protect orangutans and their habitats. A drone takes a picture every 2 seconds. How many pictures does the drone take in 30 minutes?

Review & Refresh

Find the product. Check whether your answer is reasonable.

8. Estimate: _____
   41 × 22 = _____

9. Estimate: _____
   87 × 19 = _____

10. Estimate: _____
    36 × 59 = _____

252

Name _____

Performance Task 5

The students in fourth grade go on a field trip to a planetarium.

1. The teachers have $760 to buy all of the tickets for the teachers and students. They receive less than $6 in change.

   a. Each ticket costs $6. How many tickets do the teachers buy?

   b. Exactly how much money is left over?

   c. There are 6 groups on the field trip. Each group has 1 teacher. There are an equal number of students in each group. How many students are in each group?

   d. Two groups can be in the planetarium for each show. The planetarium has 7 rows of seats with 8 seats in each row. How many seats are empty during each show?

2. The groups will be at the planetarium from 11:00 A.M. until 2:30 P.M. During that time they will rotate through 7 events: the planetarium show, 5 activities, and lunch. The planetarium show lasts 45 minutes. Each activity lasts 22 minutes. Students have 5 minutes between each event. How long does each group have to eat lunch?

3. You learn that the distance around Mars is about twice the distance around the moon. The distance around Mars is 13,263 miles. To find the distance around the moon, do you think an estimate or an exact answer is needed? Explain.

Chapter 5

Division Dots

Directions:

1. Players take turns connecting two dots, each using a different color.
2. On your turn, connect two dots, vertically or horizontally. If you close a square around a division problem, find and write the quotient and the remainder. If you do not close a square, your turn is over.
3. Continue playing until all division problems are solved.
4. The player with the most completed squares wins!

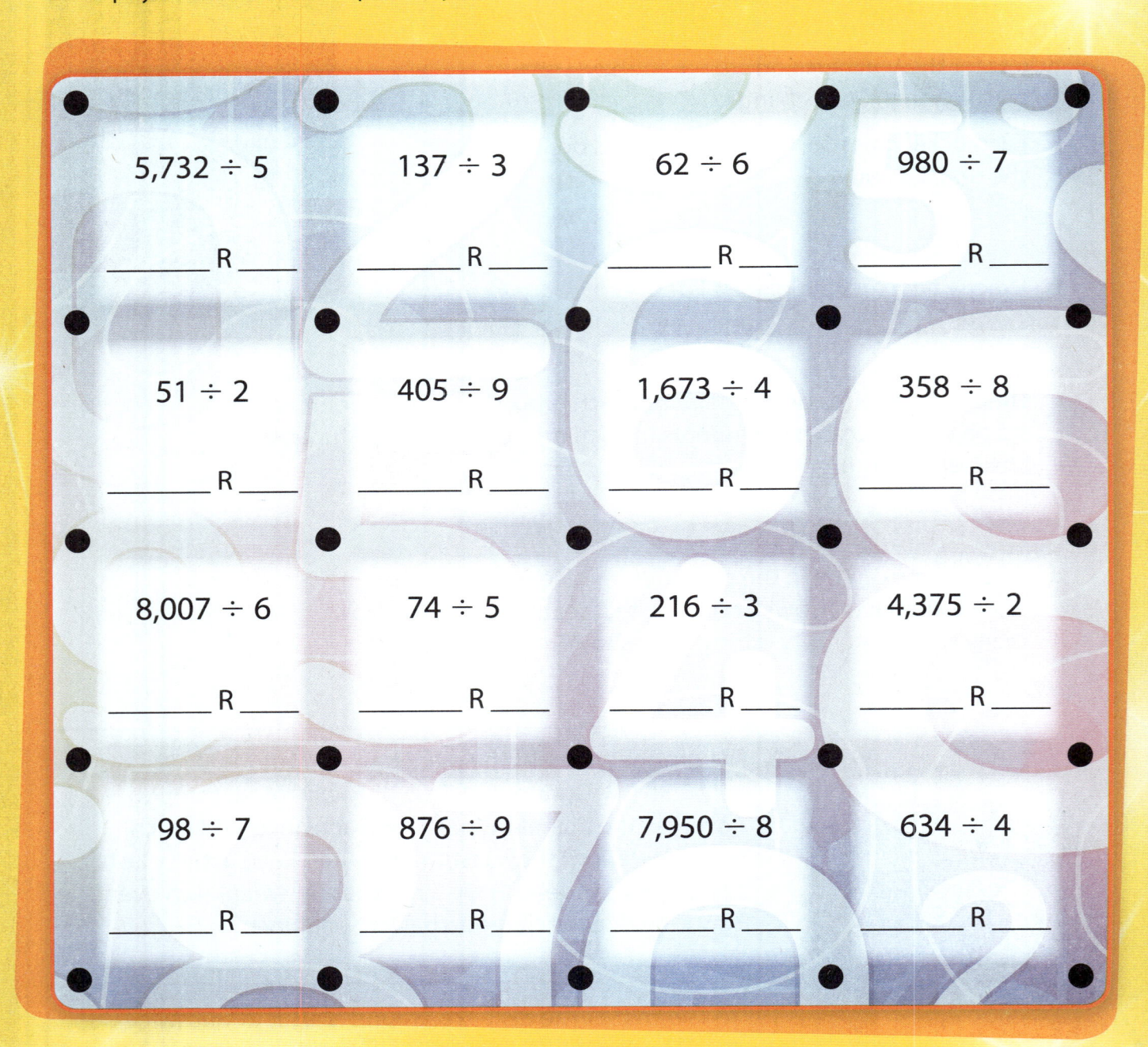

5,732 ÷ 5 ____ R ____	137 ÷ 3 ____ R ____	62 ÷ 6 ____ R ____	980 ÷ 7 ____ R ____
51 ÷ 2 ____ R ____	405 ÷ 9 ____ R ____	1,673 ÷ 4 ____ R ____	358 ÷ 8 ____ R ____
8,007 ÷ 6 ____ R ____	74 ÷ 5 ____ R ____	216 ÷ 3 ____ R ____	4,375 ÷ 2 ____ R ____
98 ÷ 7 ____ R ____	876 ÷ 9 ____ R ____	7,950 ÷ 8 ____ R ____	634 ÷ 4 ____ R ____

6 Factors, Multiples, and Patterns

- Have you ever watched a basketball game? What was your favorite part?
- An equal number of fans can sit in each bleacher row. How can you use factors to determine the total number of fans that can sit in the bleachers?

Chapter Learning Target:
Understand factors, multiples, and patterns.

Chapter Success Criteria:
- I can find the factors of a number.
- I can explain the differences between factors and multiples.
- I can compare the different features of different numbers and shapes.
- I can apply an appropriate strategy to show relationships in numbers and shapes.

6 Vocabulary

Name _____

Review Words
factors
product

Organize It

Complete the graphic organizer.

[_____]

Numbers that are multiplied to get a [_____]

____ × ____ = 18

____ × ____ = 24

Define It

Use your vocabulary cards to identify the word. Find the word in the word search.

1. Two factors that, when multiplied, result in a given product

2. A number is _____ by another number when the quotient is a whole number and the remainder is 0.

3. Tells how numbers or shapes in a pattern are related

```
F  B  I  L  T  D  O  F  R  C
H  A  E  R  I  U  Y  P  E  A
V  Q  C  W  P  E  I  J  L  R
O  U  B  T  S  G  A  D  B  U
R  M  I  L  O  N  B  N  I  M
S  L  A  E  Z  R  L  V  S  T
B  C  F  K  P  D  P  T  I  E
D  E  R  U  L  E  C  A  V  O
O  I  K  V  W  B  U  L  I  F
N  T  P  E  H  C  M  X  D  R
```

260

Chapter 6 Vocabulary Cards

composite number

divisible

factor pair

multiple

prime number

rule

A number is divisible by another number when the quotient is a whole number and the remainder is 0. $48 \div 4 = 12$ R0 So, 48 is divisible by 4.	A whole number greater than 1 with more than two factors 27 The factors of 27 are 1, 3, 9, and 27.
The product of a number and any other counting number. $1 \times 4 = 4$ $2 \times 4 = 8$ $3 \times 4 = 12$ $4 \times 4 = 16$ multiples of 4	Two factors that, when multiplied, result in a given product factor pair $2 \times 4 = 8$ factor factor 2 and 4 are a factor pair for 8.
Tells how numbers or shapes in a pattern are related Rule: Add 3. 3, 6, 9, 12, 15, 18, 21, 24, . . . Rule: triangle, hexagon, square, rhombus	A number greater than 1 with exactly two factors, 1 and itself 11 The factors of 11 are 1 and 11.

Name _____

Understand Factors 6.1

Learning Target: Use models to find factor pairs.
Success Criteria:
- I can draw area models that show a product.
- I can find the factors of a number.
- I can find the factor pairs for a number.

Explore and Grow

Draw two different rectangles that each have an area of 24 square units. Label their side lengths.

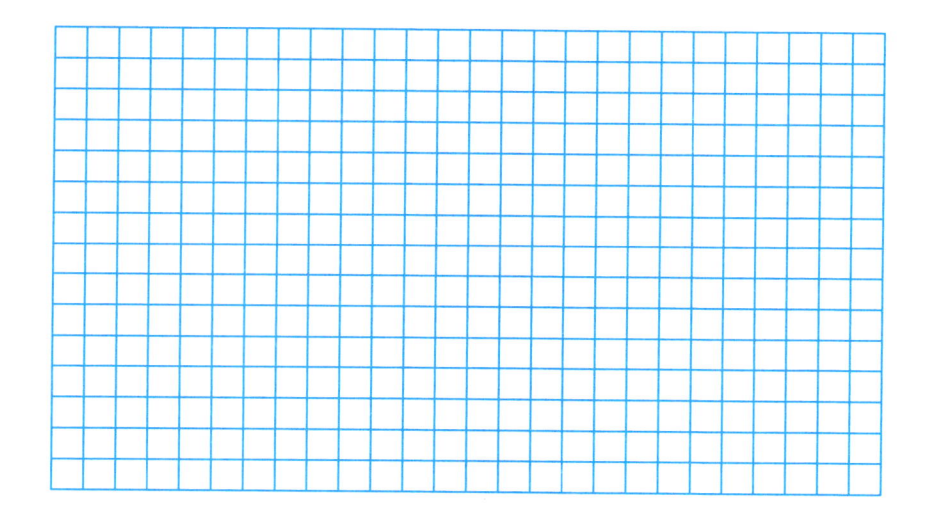

Compare your rectangles to your partner's rectangles. How are they the same? How are they different?

MP Structure How is each side length related to 24?

Chapter 6 | Lesson 1   261

Think and Grow: Find Factor Pairs

You can write whole numbers as products of two factors. The two factors are called a **factor pair** for the number.

$$2 \times 4 = 8$$

factor pair: 2 and 4
factors: 2, 4

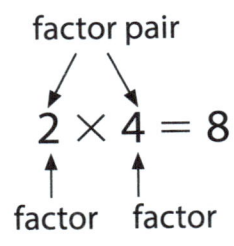

2 and 4 are a factor pair for 8.

Example Find the factor pairs for 20.

Find the side lengths of as many different rectangles with an area of 20 square units as possible.

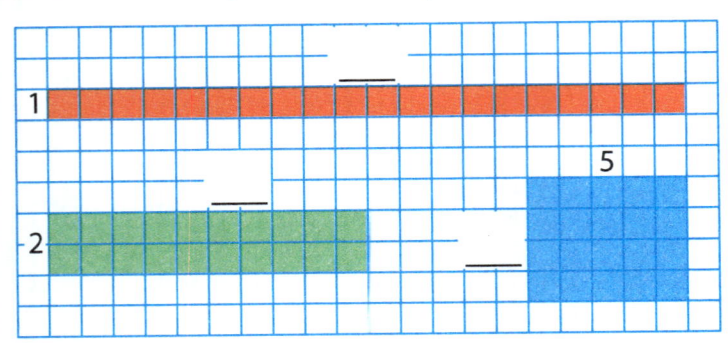

A 4 × 5 rectangle has the same area as a 5 × 4 rectangle. Both give the factor pair 4 and 5.

The side lengths of each rectangle are a factor pair.

So, the factor pairs for 20 are ____ and ____, ____ and ____, and ____ and ____.

Show and Grow   I can do it!

1. Use the rectangles to find the factor pairs for 12.

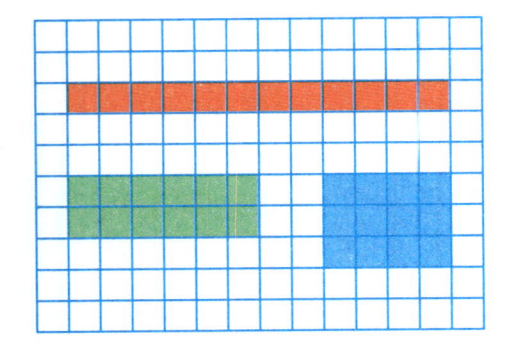

2. Draw rectangles to find the factor pairs for 16.

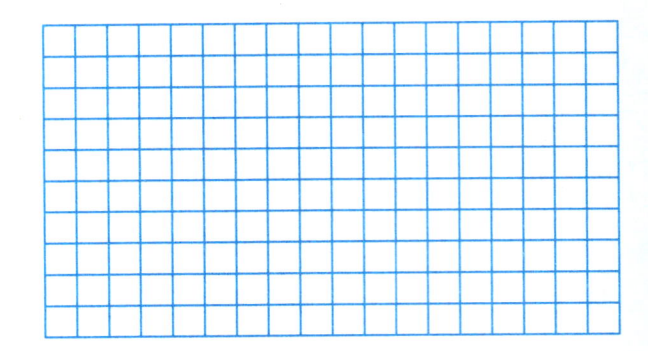

Name _____

Factors and Divisibility 6.2

Learning Target: Use division to find factor pairs.
Success Criteria:
- I can divide to find factor pairs.
- I can use divisibility rules to find factor pairs.

Explore and Grow

List any 10 multiples of 3. What do you notice about the sum of the digits in each multiple?

List any 10 multiples of 9. What do you notice about the sum of the digits in each multiple?

MP **Structure** How can you use your observations above to determine whether 3 and 9 are factors of a given number? Explain.

Chapter 6 | Lesson 2

Think and Grow: Find Factors and Factor Pairs

A number is **divisible** by another number when the quotient is a whole number and the remainder is 0.

Some numbers have divisibility rules that you can use to determine whether they are factors of other numbers.

Divisor	Divisibility Rule
2	The number is even.
3	The sum of the digits is divisible by 3.
5	The ones digit is 0 or 5.
6	The number is even and divisible by 3.
9	The sum of the digits is divisible by 9.
10	The ones digit is 0.

Example Find the factor pairs for 48.

A number is divisible by its factors.

Use divisibility rules and division to find the factors of 48.

Divisor	Is the number a factor of 48?	Multiplication Equation
1	Yes, 1 is a factor of every number.	1 × _____ = 48
2	_____, 48 is even.	2 × _____ = 48
3	_____, 4 + 8 = 12 is divisible by 3.	3 × _____ = 48
4	_____, 48 ÷ 4 = 12 R0	4 × _____ = 48
5	_____, the ones digit is not 0 or 5.	
6	_____, 48 is even and divisible by 3.	6 × _____ = 48
7	_____, 48 ÷ 7 = 6 R6	
8	_____, 48 ÷ 8 = 6 R0	8 × _____ = 48

You can stop after checking 8 because the factor pairs start to repeat.

The factors of 48 are _____, _____, _____, _____, _____, _____, _____, _____, _____, and _____.

The factor pairs for 48 are _____.

Show and Grow   I can do it!

Find the factor pairs for the number.

1. 30

2. 54

Name _____

Apply and Grow: Practice

Find the factor pairs for the number.

3. 29

4. 50

5. 63

6. 33

7. 60

8. 64

List the factors of the number.

9. 39

10. 44

11. 72

12. 67

13. 42

14. 28

15. **Reasoning** Can an odd number have an even factor? Explain.

16. **Writing** Use the diagram to explain why you do *not* have to check whether any numbers greater than 4 are factors of 12.

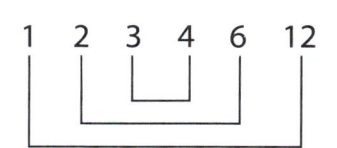

1 2 3 4 6 12

Chapter 6 | **Lesson 2** 269

Think and Grow: Modeling Real Life

Example There are 4 classes going on a field trip. The classes will use 3 buses. Can the teachers have an equal number of students on each bus?

Class	Students
1	24
2	23
3	25
4	20

Think: What do you know? What do you need to find? How will you solve?

Step 1: Add to find how many students are going on the field trip.

24 + 23 + 25 + 20 = _____

_____ students are going on the field trip.

Step 2: Is the total number of students divisible by the number of buses?

Find the sum of the digits of 92.   _____ + _____ = _____

The sum of the digits _____ divisible by 3.

The teachers _____ have an equal number of students on each bus.

Show and Grow   I can think deeper!

17. A teacher is making a 5-page test with 28 vocabulary problems and 7 reading problems. Can the teacher put an equal number of problems on each page?

18. A relay race is 39 laps long. Each team member must bike the same number of laps. Could a team have 8, 6, or 3 members? Explain.

19. **DIG DEEPER!** You have 63 clay figures to display on 7 shelves. Not all of the shelves need to be used and each shelf can hold no more than 25 figures. Each shelf must have the same number of figures. What are all the ways you could arrange the figures?

Name _____

Homework & Practice 6.2

Learning Target: Use division to find factor pairs.

Example Find the factor pairs for 90.

Use divisibility rules and division to find the factors of 90.

Divisor	Is the number a factor of 90?	Multiplication Equation
1	Yes, 1 is a factor of every number.	1 × __90__ = 90
2	__Yes__, 90 is even.	2 × __45__ = 90
3	__Yes__, 9 + 0 = 9 is divisible by 3.	3 × __30__ = 90
4	__No__, 90 ÷ 4 = 22 R2	
5	__Yes__, the ones digit is 0 or 5.	5 × __18__ = 90
6	__Yes__, 90 is even and divisible by 3.	6 × __15__ = 90
7	__No__, 90 ÷ 7 = 12 R6	
8	__No__, 90 ÷ 8 = 11 R2	
9	__Yes__, 9 + 0 = 9 is divisible by 9.	9 × __10__ = 90

The factors of 90 are __1__, __2__, __3__, __5__, __6__, __9__, __10__, __15__, __18__, __30__, __45__, and __90__.

The factor pairs for 90 are __1 and 90, 2 and 45, 3 and 30, 5 and 18, 6 and 15, and 9 and 10__.

Find the factor pairs for the number.

1. 24

2. 48

3. 31

4. 99

5. 45

6. 26

Chapter 6 | Lesson 2   271

List the factors of the number.

7. 25

8. 56

9. 75

10. 80

11. 93

12. 61

13. **Reasoning** Why does a number that has 9 as a factor also have 3 as a factor?

14. **DIG DEEPER!** The number below has 3 as a factor. What could the unknown digit be?

3 ___ 5

15. **Number Sense** Which numbers have 5 as a factor?

50   34   25   1,485   100   48

16. **Modeling Real Life** You and a partner are conducting a bottle flipping experiment. You have 3 bottles with different amounts of water in each. You need to flip each bottle 15 times. If you take turns, will you and your partner each get the same number of flips?

17. **Modeling Real Life** A florist has 55 flowers. She wants to put the same number of flowers in each vase without any left over. Should she put 2, 3, or 5 flowers in each vase? Explain.

Review & Refresh

Compare.

18. 7,914 ◯ 7,912

19. 65,901 ◯ 67,904

20. 839,275 ◯ 839,275

272

Relate Factors and Multiples

6.3

Learning Target: Understand the relationship between factors and multiples.

Success Criteria:
- I can tell whether a number is a multiple of another number.
- I can tell whether a number is a factor of another number.
- I can explain the relationship between factors and multiples.

Explore and Grow

List all factors of 24.

List several multiples of each factor. What number appears in each list?

MP Number Sense How are factors and multiples related?

Chapter 6 | Lesson 3

273

Think and Grow: Identify Multiples

A whole number is a multiple of each of its factors.

12 is a multiple of 1, 2, 3, 4, 6, and 12.

$$1 \times 12 = 12$$
$$2 \times 6 = 12$$
$$3 \times 4 = 12$$

Example Is 56 a multiple of 7?

One Way: List multiples of 7.

$1 \times 7 = 7,$
$2 \times 7 = 14,$
$3 \times 7 = 21, ...$

7, 14, 21, _____, _____, _____, _____, _____

So, 56 _____ a multiple of 7.

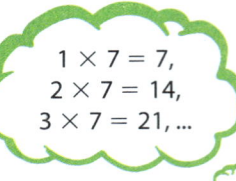

Another Way: Use division to determine whether 7 is a factor of 56.

$56 \div 7 =$ _____

7 _____ a factor of 56.

So, 56 _____ a multiple of 7.

Example Is 9 a factor of 64?

One Way: Use divisibility rules to determine whether 9 is a factor of 64.

9 _____ a factor of 64 because

$6 + 4 = 10$ _____ divisible by 9.

Another Way: List multiples of 9.

9, 18, 27, _____, _____, _____, _____, _____

64 _____ a multiple of 9.

So, 9 _____ a factor of 64.

Show and Grow   I can do it!

1. Is 23 a multiple of 3? Explain.

2. Is 8 a factor of 56? Explain.

274

Name _____

Identify Prime and Composite Numbers
6.4

Learning Target: Tell whether a given number is prime or composite.

Success Criteria:
- I can explain what prime and composite numbers are.
- I can identify prime and composite numbers.

Explore and Grow

Draw as many different rectangles as possible that each have the given area. Label their side lengths.

32 square units	13 square units

Compare the numbers of factors of 32 and 13.

Reasoning Can a whole number have fewer than two factors? exactly two factors? more than two factors?

Chapter 6 | Lesson 4

279

Think and Grow: Identify Prime and Composite Numbers

A **prime number** is a whole number greater than 1 with exactly two factors, 1 and itself. A **composite number** is a whole number greater than 1 with more than two factors.

Example Tell whether 27 is *prime* or *composite*.

Use divisibility rules.

- 27 is odd, so it _____ divisible by 2 or any other even number.

- 2 + 7 = 9 is divisible by 3, so 27 _____ divisible by 3.

Think: Odd numbers are not divisible by even numbers.

27 has factors in addition to 1 and itself.

So, 27 is _____.

Example Tell whether 11 is *prime* or *composite*.

Use divisibility rules.

- 11 is odd, so it _____ divisible by 2 or any other even number.

- 1 + 1 = 2 is not divisible by 3 or 9, so 11 _____ divisible by 3 or 9.

- The ones digit is not 0 or 5, so 11 _____ divisible by 5.

11 has exactly two factors, 1 and itself.

So, 11 is _____.

Show and Grow   I can do it!

Tell whether the number is *prime* or *composite*. Explain.

1. 7

2. 12

3. 2

4. 19

5. 45

6. 54

280

Name _____

Number Patterns 6.5

Learning Target: Create and describe number patterns.
Success Criteria:
- I can create a number pattern given a number rule.
- I can describe features of a number pattern.

Explore and Grow

Shade every third square in the table.

1	2	3	4	5	6	7	8	9	10
11	12	13	14	15	16	17	18	19	20
21	22	23	24	25	26	27	28	29	30
31	32	33	34	35	36	37	38	39	40
41	42	43	44	45	46	47	48	49	50
51	52	53	54	55	56	57	58	59	60

Write the shaded numbers. What patterns do you see?

What other patterns do you see in the table?

MP Structure Circle every fourth square in the table. Write the circled numbers. What patterns do you see?

Chapter 6 | Lesson 5   285

Think and Grow: Create Number Patterns

A **rule** tells how numbers or shapes in a pattern are related.

Example Use the rule "Add 3." to create a number pattern. The first number in the pattern is 3. Then describe another feature of the pattern.

Create the pattern.

$$+3 \quad +3 \quad +3 \quad +3 \quad +3$$

3, 6, ____, ____, ____, ____, ...

The numbers in the pattern are multiples of ____.

Example Use the rule "Multiply by 2." to create a number pattern. The first number in the pattern is 10. Then describe another feature of the pattern.

When describing another feature of the pattern, look at the ones digits or the tens digits. Are all of the numbers even or odd?

Create the pattern.

$$\times 2 \quad \times 2 \quad \times 2 \quad \times 2 \quad \times 2$$

10, 20, ____, ____, ____, ____, ...

The ones digit of each number in the pattern is ____.

Show and Grow   I can do it!

Write the first six numbers in the pattern. Then describe another feature of the pattern.

1. Rule: Add 5.
   First number: 1

   1, ____, ____, ____, ____, ____

2. Rule: Multiply by 3.
   First number: 3

   3, ____, ____, ____, ____, ____

3. Rule: Subtract 2.
   First number: 20

4. Rule: Divide by 2.
   First number: 256

Name _____

Homework & Practice 6.5

Learning Target: Create and describe number patterns.

Example Use the rule "Subtract 6." to create a number pattern. The first number in the pattern is 100. Then describe another feature of the pattern.

> When describing another feature of the pattern, look at the ones digits or the tens digits. Are all of the numbers even or odd?

Create the pattern.

100, 94, __88__, __82__, __76__, __70__, ...
(− 6 each step)

The numbers in the pattern are __even__.

Example Use the rule "Divide by 5." to create a number pattern. The first number in the pattern is 3,125. Then describe another feature of the pattern.

Create the pattern.

3,125, 625, __125__, __25__, __5__, __1__, ...
(÷ 5 each step)

The ones digit of each number in the pattern except the last number is __5__.

Write the first six numbers in the pattern. Then describe another feature of the pattern.

1. Rule: Subtract 8.
   First number: 88

   88, _____, _____, _____, _____, _____

2. Rule: Multiply by 10.
   First number: 2

   2, _____, _____, _____, _____, _____

3. Rule: Add 9.
   First number: 17

4. Rule: Divide by 2.
   First number: 1,600

Chapter 6 | Lesson 5

289

Open-Ended Use the rule to generate a pattern of four numbers.

5. Rule: Divide by 5.

6. Rule: Add 8.

7. Rule: Multiply by 9.

8. Rule: Subtract 3.

9. **MP Structure** List the first ten multiples of 9. What patterns do you notice with the digits in the ones place? in the tens place?

Does this pattern continue beyond the tenth number in the pattern?

10. Modeling Real Life It takes the moon about 28 days to orbit Earth. How many times will the moon orbit Earth in 1 year?

11. DIG DEEPER! In each level of a video game, you can earn up to 10 points and lose up to 3 points. Your friend earns 9 points in the first level. If he earns and loses the maximum number of points each level, how many total points will he have after level 6?

Review & Refresh

Find the product.

12. 14 × 23 = _____

13. 48 × 60 = _____

14. 55 × 31 = _____

Name _____

Shape Patterns 6.6

Learning Target: Create and describe shape patterns.
Success Criteria:
- I can create a shape pattern given a rule.
- I can find the shape at a given position in a pattern.
- I can describe features of a shape pattern.

Explore and Grow

Create a rule using 3 different shapes. Draw the first six shapes in the pattern.

What is the next shape in the pattern?

What is the 9th shape in the pattern? Explain.

What is the 99th shape? 1,000th shape? Explain.

MP **Structure** You want to show the first 40 shapes in the pattern above. Without modeling, how many of each shape do you think you will need?

Chapter 6 | Lesson 6

291

Think and Grow: Create Shape Patterns

Example Create a shape pattern by repeating the rule "triangle, hexagon, square, rhombus." What is the 42nd shape in the pattern?

Create the pattern.

42 ÷ 4 is 10 R2, so when the pattern repeats 10 times, the 40th shape is a _____. So, the 41st shape is a _____ and the 42nd shape is a _____.

> You divide by 4 because there are four shapes in the rule.

Example Describe the dot pattern. How many dots are in the 25th figure?

Figure 1   Figure 2   Figure 3

Figure 1 has 1 column of 4 dots, so it has 1 × _____ = _____ dots.

Figure 2 has 2 columns of 4 dots, so it has 2 × _____ = _____ dots.

Figure 3 has 3 columns of 4 dots, so it has 3 × _____ = _____ dots.

The 25th figure has _____ columns of 4 dots, so it has _____ × _____ = _____ dots.

Show and Grow   I can do it!

1. Extend the pattern of shapes by repeating the rule "square, trapezoid, triangle, hexagon, triangle." What is the 108th shape in the pattern?

2. Describe the dot pattern. How many dots are in the 76th figure?

Figure 1   Figure 2   Figure 3

Name _____

✓ Apply and Grow: Practice

3. Extend the pattern of shapes by repeating the rule "oval, triangle." What is the 55th shape?

oval triangle ____ ____ ____ ____ ____ ____ ____ ____ ...

4. Extend the pattern of symbols by repeating the rule "add, subtract, multiply, divide." What is the 103rd symbol?

+ − × ÷ ____ ____ ____ ____ ____ ____ ...

5. Describe the pattern. How many squares are in the 24th figure?

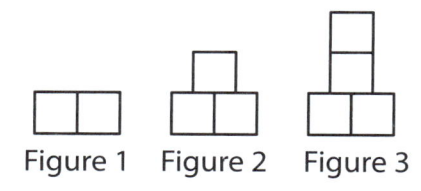

Figure 1   Figure 2   Figure 3

6. Describe the pattern of the small triangles. How many small triangles are in the 10th figure?

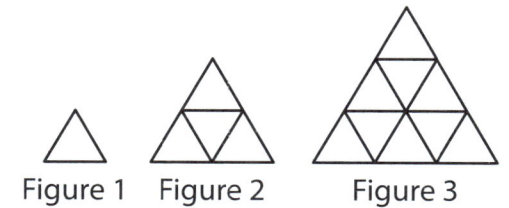

Figure 1   Figure 2   Figure 3

7. **MP Structure** Make a shape pattern that uses twice as many squares as triangles.

8. **MP Number Sense** Which shape patterns have a heart as the 12th shape?

Chapter 6 | Lesson 6   293

Think and Grow: Modeling Real Life

Example You make a necklace with cube, hexagon, and star beads. You string the beads in a pattern. You use the rule "cube, star, cube, hexagon." It takes 64 beads to complete the necklace. How many times do you repeat the pattern?

Divide the number of beads it takes to complete the necklace by the number of beads in the rule. There are 4 beads in the rule.

$$4\overline{)64}$$

You repeat the pattern _____ times.

Show and Grow   I can think deeper!

9. The path on a board game uses the rule "red, green, pink, yellow, blue." There are 55 spaces on the game board. How many times does the pattern repeat?

10. You make a walkway in a garden using different-shaped stepping stones. You use the rule "square, circle, square, hexagon." You use 24 square stepping stones. How many circle and hexagon stepping stones do you use altogether? How many stones do you use in all?

11. **DIG DEEPER!** You make a rectangular picture frame using square tiles. The picture frame is 12 tiles long and 8 tiles wide. You arrange the tiles in a pattern. You use the rule "red, orange, yellow." How many of each color tile do you use?

294

Name _____

Homework & Practice 6.6

Learning Target: Create and describe shape patterns.

Example Create a shape pattern by repeating the rule "hexagon, octagon." What is the 29th shape in the pattern?

Create the pattern.

29 divided by 2 is 14 R1, so when the pattern repeats 14 times the 28th shape is an __octagon__. So, the 29th shape is a __hexagon__.

Example Describe the dot pattern. How many dots are in the 64th figure?

Figure 1   Figure 2   Figure 3

Figure 1 has 1 row of 5 dots, so it has 1 × __5__ = __5__ dots.

Figure 2 has 2 rows of 5 dots, so it has 2 × __5__ = __10__ dots.

Figure 3 has 3 rows of 5 dots, so it has 3 × __5__ = __15__ dots.

The 64th figure has __64__ rows of 5 dots, so it has __64__ × __5__ = __320__ dots.

1. Extend the pattern of shapes by repeating the rule "up, right, down, left." What is the 48th shape in the pattern?

   ⬆ ➡ ⬇ ⬅ ____ ____ ____ ____ ____ ____ ...

2. Extend the pattern of shapes by repeating the rule "small circle, medium circle, large circle." What is the 86th shape in the pattern?

   ● ● ● ____ ____ ____ ____ ____ ____ ____ ...

Chapter 6 | Lesson 6   295

3. Describe the dot pattern. How many dots are in the 113th figure?

Figure 1   Figure 2   Figure 3

4. YOU BE THE TEACHER You and your friend each create a shape pattern with 100 shapes. Your friend says both patterns will have the same number of circles. Is your friend correct? Explain.

Friend

You

5. MP Structure Draw the missing figure in the pattern. Explain the pattern.

Figure 1   Figure 2   Figure 3   Figure 4   Figure 5

6. MP Reasoning Newton uses the rule "bone, bone, paw print" to make a shape pattern. He wants the pattern to repeat 8 times. How many bones will be in Newton's pattern?

7. Modeling Real Life The black keys on a piano follow the pattern "two black keys, three black keys." There are 36 black keys on a standard piano. How many times does this entire pattern repeat?

Review & Refresh

Find the quotient.

8. 30 ÷ 5 = _____

9. 360 ÷ 9 = _____

10. 6,400 ÷ 8 = _____

11. 140 ÷ 2 = _____

12. 4,200 ÷ 7 = _____

13. 40 ÷ 2 = _____

Name _____

Performance Task 6

You play basketball in a youth basketball program.

1. There are 72 players in the program. Each team needs an equal number of players and must have at least 5 players. What are two different ways the teams can be made?

2. The width of a basketball court is 42 feet and the length is 74 feet. You run around the perimeter of the court 4 times to warm up. How many feet do you run?

3. You and your friend are on the same team. You played your first game last week.

   a. Your team scored 4 more points than the other team. The total number of points scored by both teams was 58. How many points did your team score?

   b. You and your friend scored the same number of points. You made 2-point shots and your friend made 3-point shots. What could be the greatest number of points you and your friend each scored?

4. Your team uses the pattern below to decide which jersey color to wear to each game. Which color jersey will your team wear on the 20th game?

   Game 1   Game 2   Game 3   Game 4

Chapter 6   297

Multiple Lineup

Directions:

1. Players take turns rolling a die.
2. On your turn, place a counter on a multiple of the number of your roll. If there is not a multiple of the number of your roll, you lose your turn.
3. The first player to create a line of 5 in a row, horizontally, vertically, or diagonally, wins!

30	18	9	16	36
15	4	10	44	17
25	42	7	80	21
6	75	22	45	56
11	27	12	95	24

Name _____

Chapter Practice 6

6.1 Understand Factors

1. Use the rectangles to find the factor pairs for 6.

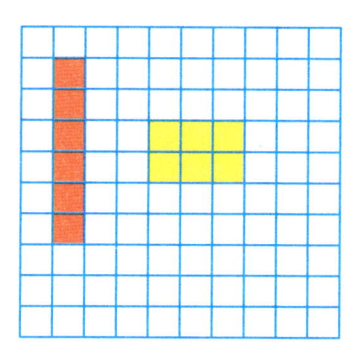

2. Draw rectangles to find the factor pairs for 12.

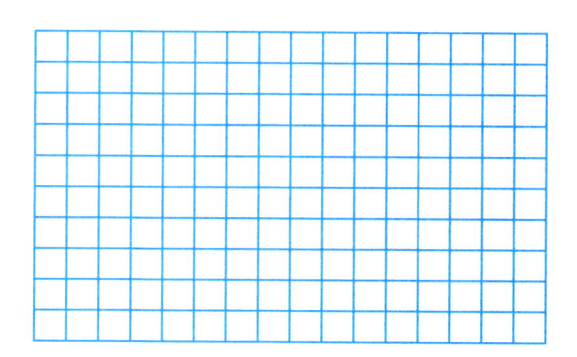

Find the factor pairs for the number.

3. 17

4. 10

5. 21

6. 20

7. 36

8. 50

6.2 Factors and Divisibility

Find the factor pairs for the number.

9. 16

10. 24

11. 56

Chapter 6   299

List the factors of the number.

12. 25

13. 60

14. 72

15. **Number Sense** Which numbers have 3 as a factor?

56   21   36   48   93   71

6.3 Relate Factors and Multiples

16. Is 54 a multiple of 3? Explain.

17. Is 45 a multiple of 7? Explain.

18. Is 2 a factor of 97? Explain.

19. Is 5 a factor of 60? Explain.

Tell whether 20 is a multiple or a factor of the number. Write *multiple*, *factor*, or *both*.

20. 60

21. 4

22. 20

23. **Number Sense** Name two numbers that are each a multiple of both 5 and 2. What do you notice about the two multiples?

24. **Logic** A quotient is a multiple of 5. The dividend is a multiple of 4. The divisor is a factor of 8. Write one possible equation for the problem.

300

6.4 Identify Prime and Composite Numbers

Tell whether the number is *prime* or *composite*. Explain.

25. 5

26. 25

27. 51

28. 21

29. 50

30. 83

31. Modeling Real Life A prime number of students have which type of fingerprint?

Loop   Arch   Whorl

Type of Fingerprint

Loop	◯◯◯◯◐
Arch	◯◯◯◯◯◯
Whorl	◯◯◯◐

Each ◯ = 2 students.

6.5 Number Patterns

Write the first six numbers in the pattern. Then describe another feature of the pattern.

32. Rule: Subtract 11.
First number: 99

99, ____, ____, ____, ____, ____

33. Rule: Multiply by 5.
First number: 10

10, ____, ____, ____, ____, ____

34. Rule: Add 8.
First number: 15

35. Rule: Divide by 4.
First number: 4,096

Open-Ended Use the rule to generate a pattern of four numbers.

36. Rule: Divide by 2.

37. Rule: Add 3.

38. Rule: Multiply by 10.

39. Rule: Subtract 6.

6.6 Shape Patterns

40. Extend the pattern of shapes by repeating the rule "trapezoid, circle." What is the 57th shape in the pattern?

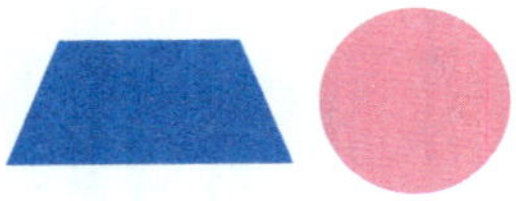

41. Extend the pattern of shapes by repeating the rule "top left, top right, bottom right, bottom left." What is the 102nd shape in the pattern?

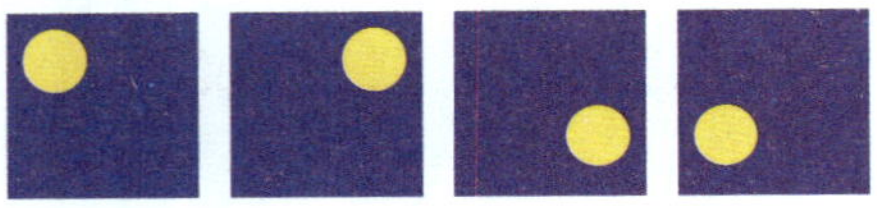

42. Describe the pattern. How many squares are in the 61st figure?

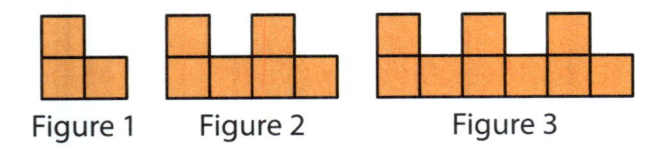

Figure 1   Figure 2   Figure 3

43. **MP Structure** Draw the missing figure in the pattern. Explain the pattern.

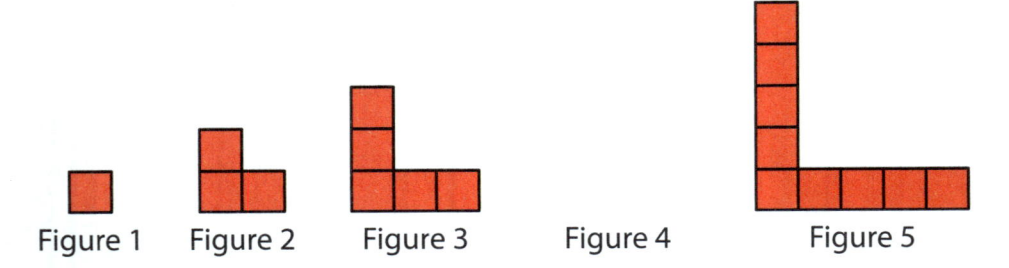

Figure 1   Figure 2   Figure 3   Figure 4   Figure 5

302

7 Understand Fraction Equivalence and Comparison

- What are your favorite colors?
- How can you use fractions to compare the amounts of each color of paint you use?

Chapter Learning Target:
Understand fractions.

Chapter Success Criteria:
- I can define equivalent fractions.
- I can explain how multiplication can be used to find equivalent fractions.
- I can compare the numerators and denominators of two fractions.
- I can find the factors of a number.

Name _____

7 Vocabulary

Review Words
denominator
fraction
numerator

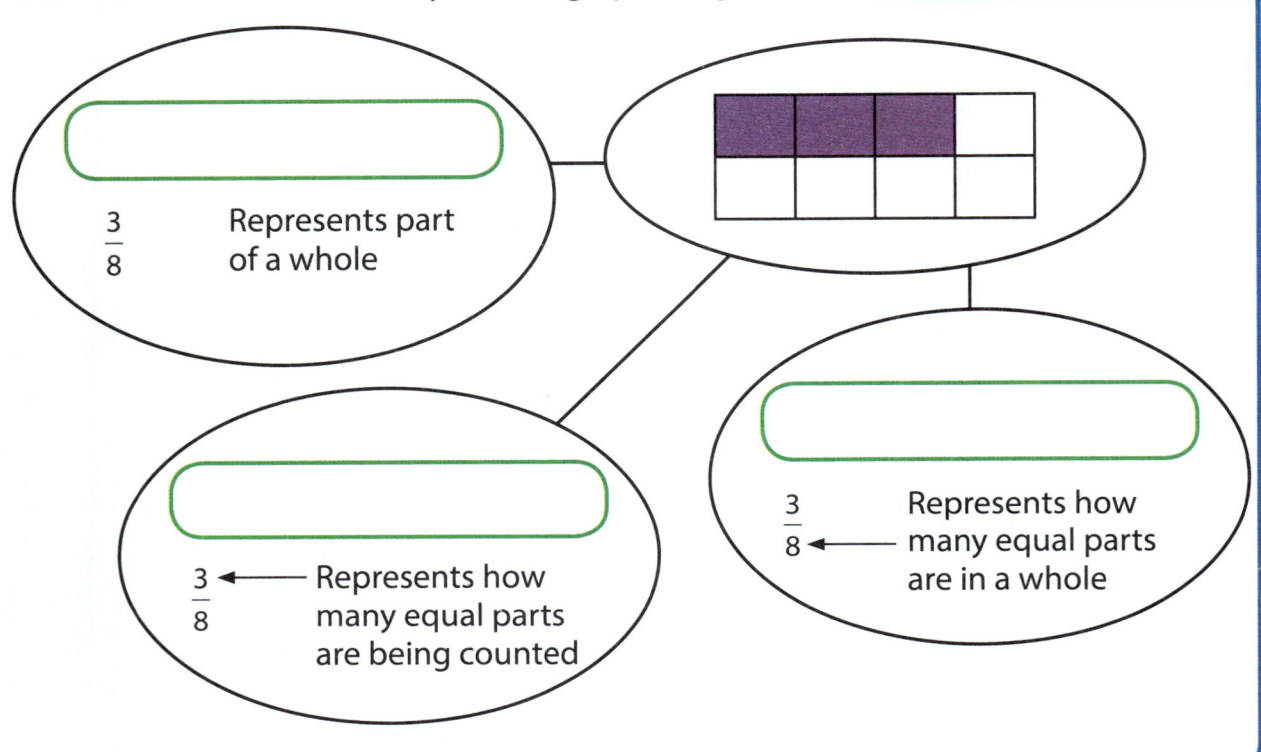

Organize It

Use the review words to complete the graphic organizer.

(_____)
$\frac{3}{8}$ Represents part of a whole

(_____)
$\frac{3}{8}$ ← Represents how many equal parts are being counted

(_____)
$\frac{3}{8}$ ← Represents how many equal parts are in a whole

Define It

Use your vocabulary cards to complete each definition.

1. benchmark: A commonly used _____ that you can use to _____ other numbers

2. common factor: A _____ that is _____ by two or more given _____

3. equivalent fractions: Two or more _____ that name the _____ part of a _____

304

Chapter 7 Vocabulary Cards

benchmark

common factor

equivalent

equivalent fractions

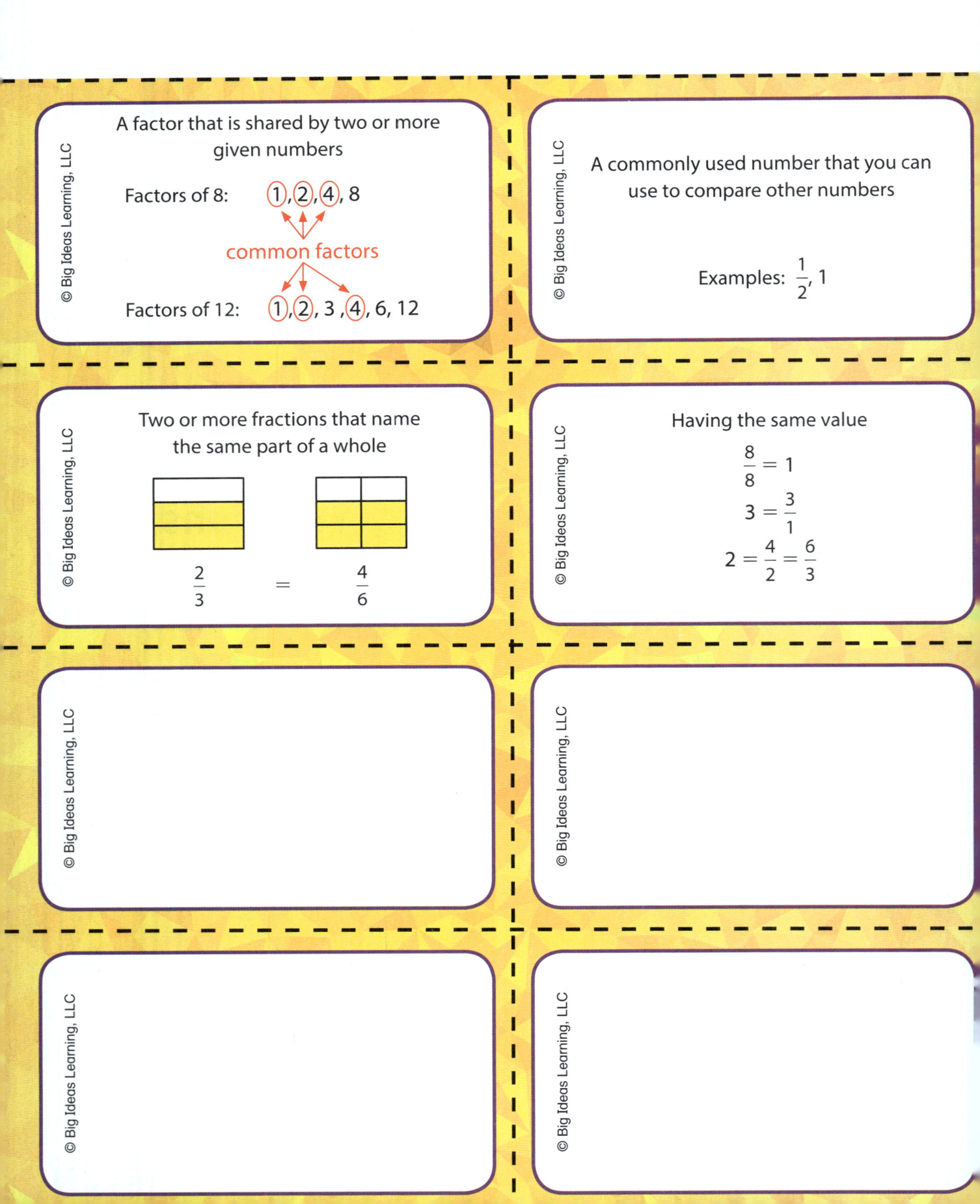

A factor that is shared by two or more given numbers Factors of 8: ①,②,④, 8 common factors Factors of 12: ①,②, 3 ,④, 6, 12	A commonly used number that you can use to compare other numbers Examples: $\frac{1}{2}$, 1
Two or more fractions that name the same part of a whole $\frac{2}{3} = \frac{4}{6}$	Having the same value $\frac{8}{8} = 1$ $3 = \frac{3}{1}$ $2 = \frac{4}{2} = \frac{6}{3}$

Name _____

Model Equivalent Fractions 7.1

Learning Target: Model and write equivalent fractions.

Success Criteria:
- I can use an area model to find equivalent fractions.
- I can use a number line to find equivalent fractions.
- I can write equivalent fractions.

Explore and Grow

Use the model to write fractions that are the same size as $\frac{1}{2}$.

1 whole												
$\frac{1}{2}$						$\frac{1}{2}$					
$\frac{1}{3}$				$\frac{1}{3}$				$\frac{1}{3}$			
$\frac{1}{4}$			$\frac{1}{4}$			$\frac{1}{4}$			$\frac{1}{4}$		
$\frac{1}{5}$			$\frac{1}{5}$		$\frac{1}{5}$			$\frac{1}{5}$		$\frac{1}{5}$		
$\frac{1}{6}$		$\frac{1}{6}$		$\frac{1}{6}$		$\frac{1}{6}$		$\frac{1}{6}$		$\frac{1}{6}$	
$\frac{1}{8}$	$\frac{1}{8}$	$\frac{1}{8}$	$\frac{1}{8}$		$\frac{1}{8}$	$\frac{1}{8}$	$\frac{1}{8}$	$\frac{1}{8}$		
$\frac{1}{10}$	$\frac{1}{10}$	$\frac{1}{10}$	$\frac{1}{10}$	$\frac{1}{10}$	$\frac{1}{10}$	$\frac{1}{10}$	$\frac{1}{10}$	$\frac{1}{10}$	$\frac{1}{10}$		
$\frac{1}{12}$	$\frac{1}{12}$	$\frac{1}{12}$	$\frac{1}{12}$	$\frac{1}{12}$	$\frac{1}{12}$	$\frac{1}{12}$	$\frac{1}{12}$	$\frac{1}{12}$	$\frac{1}{12}$	$\frac{1}{12}$	$\frac{1}{12}$

MP Reasoning Can you write a fraction with a denominator of 12 that is the same size as $\frac{2}{3}$? Explain.

Chapter 7 | Lesson 1   305

Think and Grow: Model Equivalent Fractions

Two or more numbers that have the same value are **equivalent**. Two or more fractions that name the same part of a whole are **equivalent fractions**. Equivalent fractions name the same point on a number line.

Example Use models to find equivalent fractions for $\frac{2}{3}$.

One Way: Draw models that show the same whole divided into different numbers of parts.

The fractions shown by the second and third models are also equivalent.

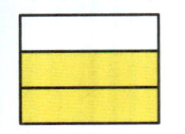

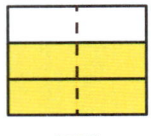

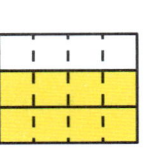

$\frac{2}{3}$ $\frac{\square}{\square}$ $\frac{\square}{\square}$

$\frac{2}{3}$ is equivalent to $\frac{\square}{\square}$ and $\frac{\square}{\square}$.

So, $\frac{2}{3} = \frac{\square}{\square}$ and $\frac{2}{3} = \frac{\square}{\square}$.

Another Way: Use a number line.

Step 1: Plot $\frac{2}{3}$ on a number line.

Step 2: Divide the number line into sixths and label the tick marks.

You can also divide the number line into twelfths.

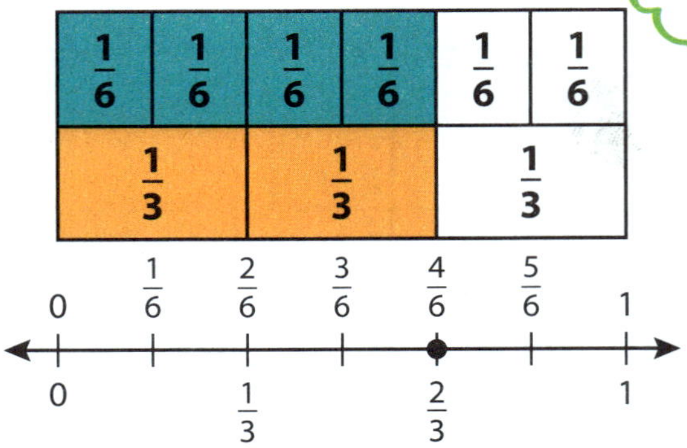

The fractions that name the same point are $\frac{2}{3}$ and $\frac{\square}{\square}$. So, $\frac{2}{3} = \frac{\square}{\square}$.

Show and Grow   I can do it!

1. Use the model to find an equivalent fraction for $\frac{2}{5}$.

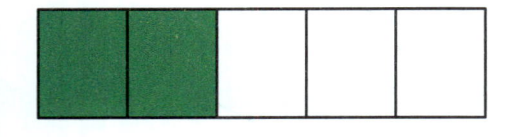

2. Use the number line to find an equivalent fraction for $\frac{1}{6}$.

306

Name _____

Learning Target: Model and write equivalent fractions.

Homework & Practice 7.1

Example Use the models to find equivalent fractions for $\frac{1}{2}$.

One Way: Draw models that show the same whole divided into different numbers of parts.

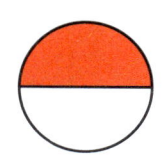

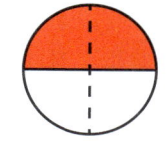

$\frac{1}{2}$ $\frac{2}{4}$ $\frac{4}{8}$

$\frac{1}{2}$ is equivalent to $\frac{2}{4}$ and $\frac{4}{8}$. So, $\frac{1}{2} = \frac{2}{4}$ and $\frac{1}{2} = \frac{4}{8}$.

Another Way: Use a number line.

Step 1: Plot $\frac{1}{2}$ on a number line.

Step 2: Divide the number line into fourths and label the tick marks.

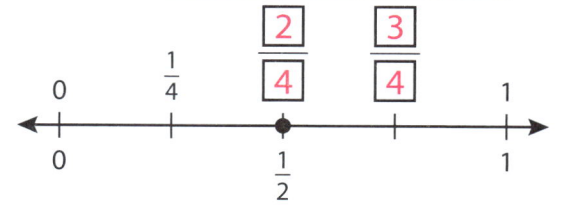

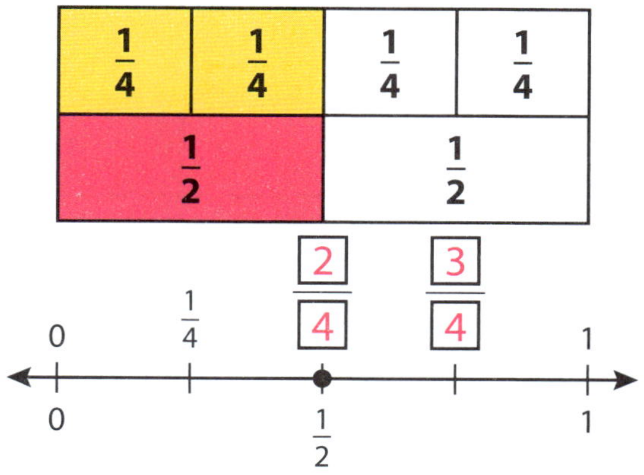

| $\frac{1}{4}$ | $\frac{1}{4}$ | $\frac{1}{4}$ | $\frac{1}{4}$ |
| $\frac{1}{2}$ | | $\frac{1}{2}$ | |

0   $\frac{1}{4}$ $\frac{2}{4}$ $\frac{3}{4}$ 1

0   $\frac{1}{2}$ 1

The fractions that name the same point are $\frac{1}{2}$ and $\frac{2}{4}$. So, $\frac{1}{2} = \frac{2}{4}$.

Use the model to find an equivalent fraction.

1. $\frac{3}{5}$

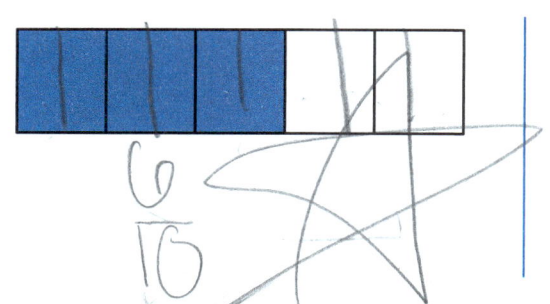

2. $\frac{1}{4}$

Chapter 7 | Lesson 1   309

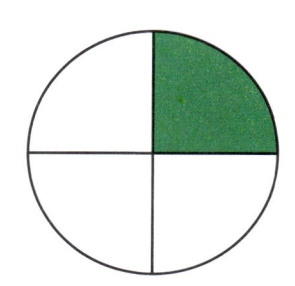

Use the number line to find an equivalent fraction.

3. $\frac{4}{6}$

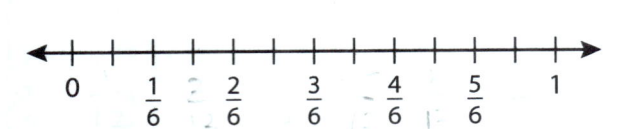

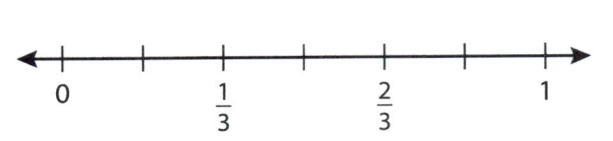

0    $\frac{1}{6}$ $\frac{2}{6}$ $\frac{3}{6}$ $\frac{4}{6}$ $\frac{5}{6}$ 1

4. $\frac{2}{3}$

0    $\frac{1}{3}$ $\frac{2}{3}$ 1

Find the equivalent fraction.

5. $\frac{1}{6} = \frac{\square}{12}$

6. $\frac{2}{5} = \frac{\square}{10}$

7. $\frac{1}{4} = \frac{\square}{8}$

8. $\frac{9}{12} = \frac{\square}{4}$

9. **Which One Doesn't Belong?** Which model does *not* belong with the other three? Explain.

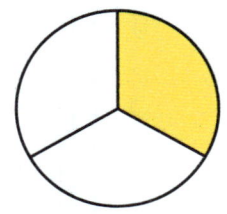

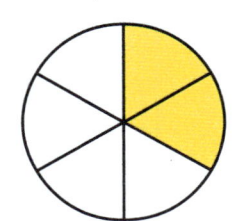

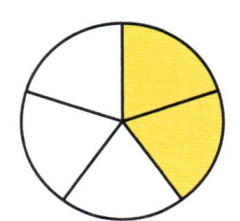

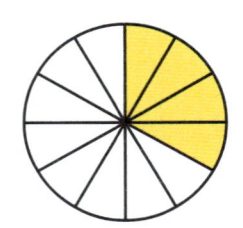

10. **Modeling Real Life** Your crayon is $\frac{1}{6}$ foot long. Your friend's crayon is $\frac{3}{12}$ foot long. Are the crayons the same length?

Review & Refresh

Find the product.

11. $5 \times 437 = $ _____

12. $6{,}982 \times 9 = $ _____

13. $8 \times 708 = $ _____

310

Name _____

Generate Equivalent Fractions by Multiplying
7.2

Learning Target: Use multiplication to find equivalent fractions.

Success Criteria:
- I can multiply a numerator and a denominator by a chosen number.
- I can multiply to find equivalent fractions.
- I can explain why multiplication can be used to find equivalent fractions.

Explore and Grow

Shade the second model in each pair to show an equivalent fraction. Then write the fraction.

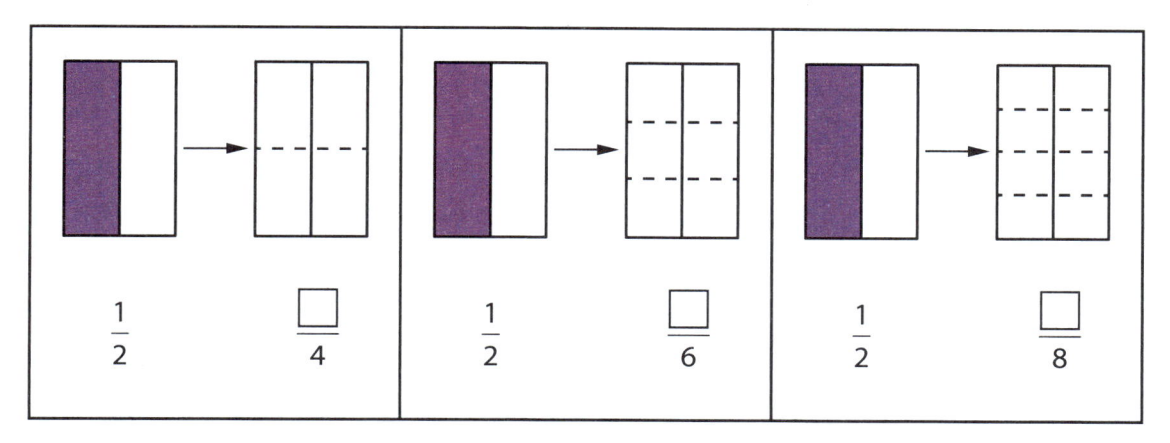

$\dfrac{1}{2}$ → $\dfrac{\square}{4}$ $\dfrac{1}{2}$ → $\dfrac{\square}{6}$ $\dfrac{1}{2}$ → $\dfrac{\square}{8}$

Describe the relationship between each pair of numerators and each pair of denominators.

MP Structure How can you use multiplication to write equivalent fractions? Explain. Then use your method to find another fraction that is equivalent to $\dfrac{1}{2}$.

Chapter 7 | Lesson 2

311

Think and Grow: Multiply to Find Equivalent Fractions

You can find an equivalent fraction by multiplying the numerator and the denominator by the same number.

$$\frac{1}{2} = \frac{1 \times 3}{2 \times 3} = \frac{3}{6}$$

Example Find an equivalent fraction for $\frac{3}{5}$.

Multiply the numerator and the denominator by 2.

$$\frac{3}{5} = \frac{3 \times 2}{5 \times 2} = \frac{\Box}{\Box}$$

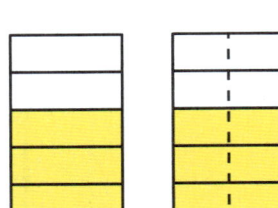

$\frac{\Box}{\Box}$ is equivalent to $\frac{3}{5}$.

$\frac{3}{5}$ is 3 parts when each part is $\frac{1}{5}$. This is the same as 6 parts when each part is $\frac{1}{10}$.

Example Find an equivalent fraction for $\frac{7}{4}$.

Multiply the numerator and the denominator by 3.

$$\frac{7}{4} = \frac{7 \times 3}{4 \times 3} = \frac{\Box}{\Box}$$

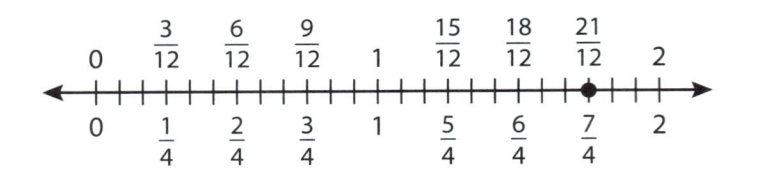

$\frac{\Box}{\Box}$ is equivalent to $\frac{7}{4}$.

Show and Grow I can do it!

Find an equivalent fraction.

1. $\frac{5}{6} = \dfrac{5 \times \Box}{6 \times \Box} = \dfrac{\Box}{\Box}$

2. $\frac{8}{5} = \dfrac{8 \times \Box}{5 \times \Box} = \dfrac{\Box}{\Box}$

Find the equivalent fraction.

3. $\frac{1}{2} = \dfrac{\Box}{8}$

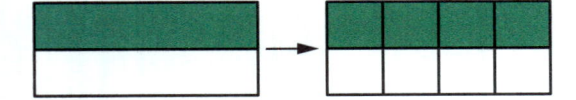

4. $\frac{2}{3} = \dfrac{\Box}{6}$

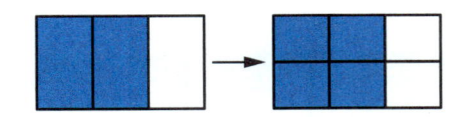

Name _____

Generate Equivalent Fractions by Dividing 7.3

Learning Target: Use division to find equivalent fractions.

Success Criteria:
- I can find the factors of a number.
- I can find the common factors of a numerator and a denominator.
- I can divide to find equivalent fractions.

Explore and Grow

Shade the second model in each pair to show an equivalent fraction. Then write the fraction.

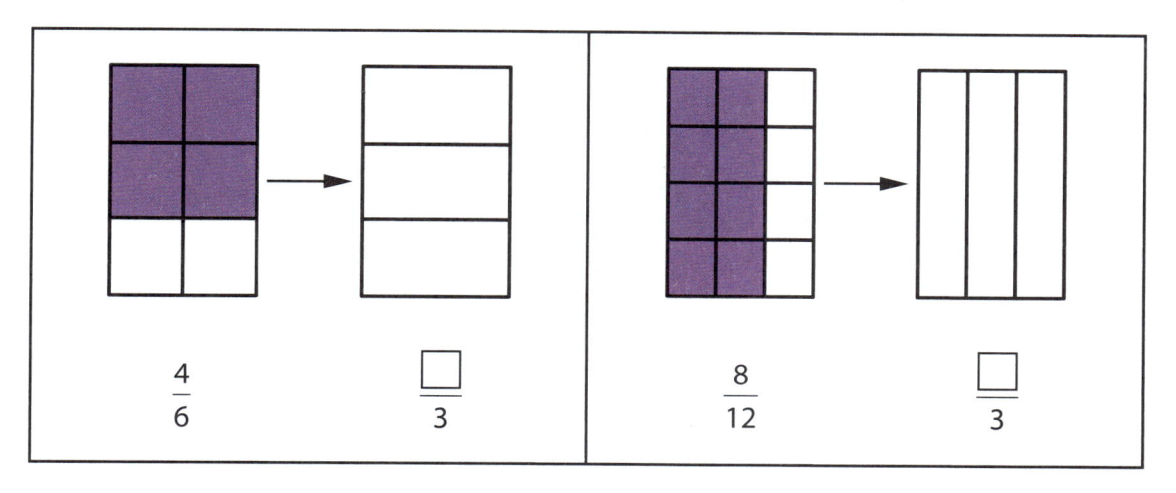

$\dfrac{4}{6}$ → $\dfrac{\square}{3}$

$\dfrac{8}{12}$ → $\dfrac{\square}{3}$

Describe the relationship between each pair of numerators and each pair of denominators.

Structure How can you use division to write equivalent fractions? Explain. Then use your method to find a fraction that is equivalent to $\dfrac{6}{10}$.

Chapter 7 | Lesson 3

Think and Grow: Divide to Find Equivalent Fractions

A factor that is shared by two or more given numbers is a **common factor**. You can find an equivalent fraction by dividing the numerator and the denominator by a common factor.

$$\frac{2}{4} = \frac{2 \div 2}{4 \div 2} = \frac{1}{2}$$

Example   Find an equivalent fraction for $\frac{8}{12}$.

Find the common factors of 8 and 12.
- The factors of 8 are ①, ②, ④, and 8.
- The factors of 12 are ①, ②, 3, ④, 6, and 12.

So, the common factors are ____, ____, and ____.

To find an equivalent fraction, divide the numerator and the denominator by the common factor 2.

$$\frac{8}{12} = \frac{8 \div \square}{12 \div \square} = \frac{\square}{\square}$$

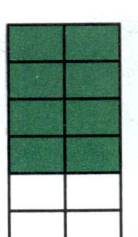

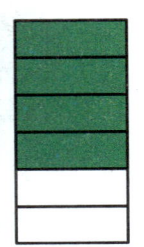

$\frac{\square}{\square}$ is equivalent to $\frac{8}{12}$.

> You can also divide the numerator and the denominator by the common factor 4 to find an equivalent fraction.

Show and Grow   I can do it!

Find an equivalent fraction.

1. $\frac{3}{6} = \frac{3 \div \square}{6 \div \square} = \frac{\square}{\square}$

2. $\frac{20}{8} = \frac{20 \div \square}{8 \div \square} = \frac{\square}{\square}$

Find the equivalent fraction.

3. $\frac{4}{10} = \frac{\square}{5}$

4. $\frac{90}{100} = \frac{9}{\square}$

5. $\frac{14}{4} = \frac{\square}{2}$

318

Name _____

Compare Fractions Using Benchmarks 7.4

Learning Target: Compare fractions using benchmarks.

Success Criteria:
- I can compare a fraction to a benchmark of $\frac{1}{2}$ or 1.
- I can use a benchmark to compare two fractions.

Explore and Grow

Use a model to compare $\frac{1}{2}$ and $\frac{5}{8}$.

$\frac{1}{2}$ ◯ $\frac{5}{8}$

Use a model to compare $\frac{1}{2}$ and $\frac{2}{5}$.

$\frac{1}{2}$ ◯ $\frac{2}{5}$

How can you use your results to compare $\frac{5}{8}$ and $\frac{2}{5}$?

MP **Structure** How does the numerator of a fraction compare to the denominator when the fraction is less than $\frac{1}{2}$? greater than $\frac{1}{2}$? equal to $\frac{1}{2}$? Explain.

Chapter 7 | Lesson 4

Think and Grow: Compare Fractions Using Benchmarks

A **benchmark** is a commonly used number that you can use to compare other numbers. You can use the benchmarks $\frac{1}{2}$ and 1 to help you compare fractions.

7 is greater than half of 10, so $\frac{7}{10}$ is greater than $\frac{1}{2}$.

Example Use fraction strips to compare $\frac{7}{10}$ and $\frac{3}{8}$.

Compare each fraction to the benchmark $\frac{1}{2}$.

$\frac{7}{10} \bigcirc \frac{1}{2}$ and $\frac{3}{8} \bigcirc \frac{1}{2}$

So, $\frac{7}{10} \bigcirc \frac{3}{8}$.

Example Use a number line to compare $\frac{5}{6}$ and $\frac{4}{3}$.

Compare each fraction to the benchmark 1.

$\frac{5}{6} \bigcirc 1$ and $\frac{4}{3} \bigcirc 1$

So, $\frac{5}{6} \bigcirc \frac{4}{3}$.

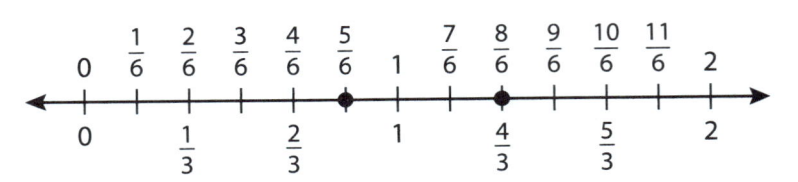

Show and Grow I can do it!

Compare. Use a model to help.

1. $\frac{5}{12} \bigcirc \frac{3}{5}$

2. $\frac{3}{4} \bigcirc \frac{6}{8}$

3. $\frac{6}{5} \bigcirc \frac{9}{10}$

Apply and Grow: Practice

Compare. Use a model to help.

4. $\frac{4}{12} \bigcirc \frac{7}{10}$

5. $\frac{1}{2} \bigcirc \frac{3}{6}$

6. $\frac{2}{10} \bigcirc \frac{5}{6}$

7. $\frac{5}{5} \bigcirc \frac{12}{12}$

8. $\frac{4}{2} \bigcirc \frac{7}{10}$

9. $\frac{4}{6} \bigcirc \frac{1}{3}$

10. $\frac{5}{4} \bigcirc \frac{3}{8}$

11. $\frac{6}{12} \bigcirc \frac{4}{5}$

12. $\frac{3}{2} \bigcirc \frac{80}{100}$

13. A black bear hibernates for $\frac{7}{12}$ of 1 year. A bat hibernates for $\frac{1}{4}$ of 1 year. Which animal hibernates longer? How do you know?

14. **Writing** Explain how you can tell whether a fraction is greater than, less than, or equal to 1, just by looking at the numerator and the denominator.

15. **DIG DEEPER!** You and your friend pack a lunch. You eat $\frac{2}{6}$ of your lunch. Your friend eats $\frac{3}{4}$ of his lunch. Can you tell who ate more? Explain.

Chapter 7 | Lesson 4   325

Think and Grow: Modeling Real Life

Example You have $\frac{3}{5}$ of a bottle of blue paint and $\frac{7}{8}$ of a bottle of yellow paint. Do you have enough of each paint color to make the recipe? Explain.

Green Paint Recipe
$\frac{2}{8}$ of a bottle of blue paint
$\frac{5}{10}$ of a bottle of yellow paint

Compare each fraction of a paint bottle to the benchmark $\frac{1}{2}$.

Blue paint: $\frac{3}{5} \bigcirc \frac{1}{2}$ and $\frac{2}{8} \bigcirc \frac{1}{2}$ So, $\frac{3}{5} \bigcirc \frac{2}{8}$.

Yellow paint: $\frac{7}{8} \bigcirc \frac{1}{2}$ and $\frac{5}{10} \bigcirc \frac{1}{2}$ So, $\frac{7}{8} \bigcirc \frac{5}{10}$.

You _____ have enough of each paint color to make the recipe.

Explain.

Show and Grow I can think deeper!

Slime Recipe
$\frac{3}{4}$ tablespoon of baking soda
$\frac{3}{2}$ tablespoons of contact lens solution
$\frac{6}{8}$ cup of glue

16. You have $\frac{3}{8}$ tablespoon of baking soda, $\frac{2}{3}$ tablespoon of contact lens solution, and $\frac{5}{3}$ cups of glue. Do you have enough of each ingredient to make the recipe? Explain.

17. **DIG DEEPER!** You and your friend are making posters for a science fair. The posters are the same size. Your poster has 8 equal parts and you are halfway done. Your friend's poster has 12 equal parts. Your friend has completed $\frac{9}{12}$ of her poster. Who has a greater amount of poster left to complete?

Name _____

Homework & Practice 7.4

Learning Target: Compare fractions using benchmarks.

Example Use fraction strips to compare $\frac{5}{12}$ and $\frac{3}{4}$.

Compare each fraction to the benchmark $\frac{1}{2}$.

$\frac{5}{12}$ ◯< $\frac{1}{2}$ and $\frac{3}{4}$ ◯> $\frac{1}{2}$

So, $\frac{5}{12}$ ◯< $\frac{3}{4}$.

Example Use a number line to compare $\frac{3}{2}$ and $\frac{7}{8}$.

Compare each fraction to the benchmark 1.

$\frac{3}{2}$ ◯> 1 and $\frac{7}{8}$ ◯< 1

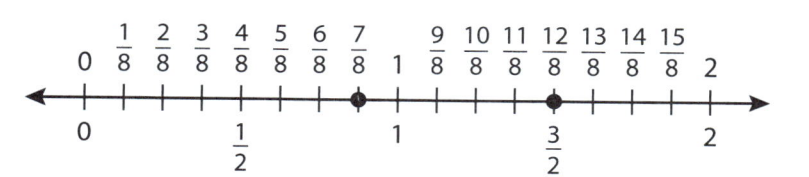

So, $\frac{3}{2}$ ◯> $\frac{7}{8}$.

Compare. Use a model to help.

1. $\frac{8}{12}$ ◯ $\frac{8}{6}$

2. $\frac{9}{10}$ ◯ $\frac{1}{3}$

3. $\frac{5}{2}$ ◯ $\frac{7}{8}$

4. $\frac{1}{4}$ ◯ $\frac{7}{12}$

5. $\frac{2}{2}$ ◯ $\frac{10}{8}$

6. $\frac{60}{100}$ ◯ $\frac{3}{5}$

7. $\frac{4}{12}$ ◯ $\frac{2}{6}$

8. $\frac{4}{6}$ ◯ $\frac{5}{100}$

9. $\frac{8}{10}$ ◯ $\frac{9}{1}$

Chapter 7 | Lesson 4

10. In a litter of kittens, $\frac{3}{4}$ are white, and $\frac{2}{8}$ are gray. Are there more white or more gray kittens?

Open-Ended Complete the statement.

11. $\frac{5}{6} < \frac{\square}{\square}$

12. $\frac{7}{8} > \frac{\square}{\square}$

13. $\frac{12}{5} > \frac{\square}{\square}$

14. **Number Sense** Which statements are true?

$\frac{8}{3} \overset{?}{>} \frac{5}{6}$ $\frac{10}{12} \overset{?}{>} \frac{4}{10}$ $\frac{2}{3} \overset{?}{<} \frac{3}{6}$ $\frac{1}{4} \overset{?}{<} \frac{2}{8}$

Smoothie Recipe
$\frac{3}{2}$ cups of oranges
$\frac{7}{8}$ cup of strawberries

15. **Modeling Real Life** You have $\frac{1}{3}$ cup of oranges and $\frac{5}{4}$ cups of strawberries. Do you have enough of each ingredient to make the smoothie? Explain.

16. **DIG DEEPER!** Newton and Descartes are picking apples at a farm. Newton's bag of apples weighs $\frac{4}{5}$ pound. Descartes's bag weighs $\frac{3}{2}$ pounds. How much money will Newton and Descartes each pay for their bag of apples?

Apple Prices
Less than $\frac{1}{2}$ pound: 50¢
$\frac{1}{2}$ pound – 1 pound: 75¢
Over 1 pound: $1

Review & Refresh

Find the factor pairs for the number.

17. 12

18. 50

19. 17

Name _____

Compare Fractions 7.5

Learning Target: Compare fractions using equivalent fractions.

Success Criteria:
- I can compare the numerators and denominators of two fractions.
- I can make the numerators or the denominators of two fractions the same.
- I can compare fractions with like numerators or like denominators.

Explore and Grow

Shade each pair of models to compare $\frac{1}{3}$ and $\frac{5}{12}$. Explain how each pair of models helps you compare the fractions differently.

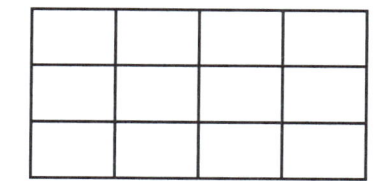

$\frac{1}{3}$ 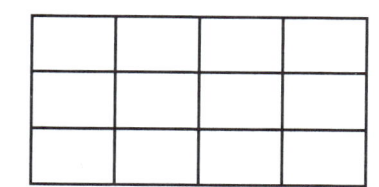 $\frac{5}{12}$

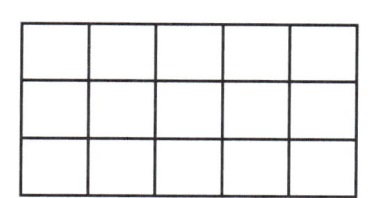

$\frac{1}{3}$ 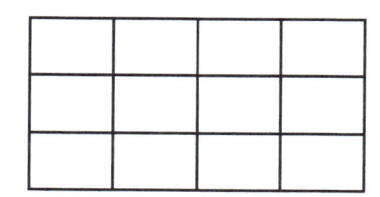 $\frac{5}{12}$

MP Reasoning How can you use equivalent fractions to compare fractions with different numerators and different denominators?

Chapter 7 | Lesson 5   329

Think and Grow: Compare Fractions

Example Compare $\frac{3}{5}$ and $\frac{9}{10}$.

One Way: Use a like denominator. Find an equivalent fraction for $\frac{3}{5}$ that has a denominator of 10.

Multiply the numerator and the denominator of $\frac{3}{5}$ by _____.

$$\frac{3}{5} = \frac{3 \times \square}{5 \times \square} = \frac{6}{10}$$

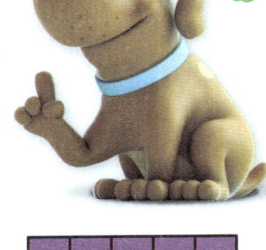

The wholes are divided into the same number of parts.

Compare $\frac{6}{10}$ and $\frac{9}{10}$.

$\frac{6}{10} \bigcirc \frac{9}{10}$

So, $\frac{3}{5} \bigcirc \frac{9}{10}$.

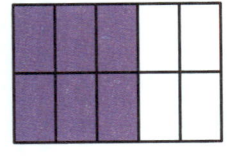

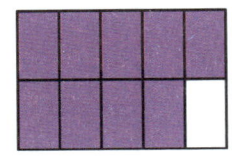

Another Way: Use a like numerator. Find an equivalent fraction for $\frac{3}{5}$ that has a numerator of 9.

Multiply the numerator and the denominator of $\frac{3}{5}$ by _____.

$$\frac{3}{5} = \frac{3 \times \square}{5 \times \square} = \frac{9}{15}$$

The wholes are divided into different of parts.

Compare $\frac{9}{15}$ and $\frac{9}{10}$.

$\frac{9}{15} \bigcirc \frac{9}{10}$

So, $\frac{3}{5} \bigcirc \frac{9}{10}$.

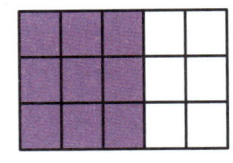

$\frac{3}{5} = \frac{6}{10}$ $\frac{9}{10}$ $\frac{3}{5} = \frac{9}{15}$  $\frac{9}{10}$

Show and Grow I can do it!

Compare. Use a model to help.

1. $\frac{7}{8} \bigcirc \frac{3}{4}$

2. $\frac{4}{6} \bigcirc \frac{2}{3}$

3. $\frac{4}{3} \bigcirc \frac{5}{4}$

Name _____

Homework & Practice 7.5

Learning Target: Compare fractions using equivalent fractions.

Example Compare $\frac{3}{4}$ and $\frac{6}{8}$.

One Way: Use a like denominator. Find an equivalent fraction for $\frac{3}{4}$ that has a denominator of 8.

Multiply the numerator and the denominator of $\frac{3}{4}$ by __2__.

$$\frac{3}{4} = \frac{3 \times \boxed{2}}{4 \times \boxed{2}} = \frac{6}{8}$$

Compare $\frac{6}{8}$ and $\frac{6}{8}$.

$\frac{6}{8}$ ⊜ $\frac{6}{8}$

So, $\frac{3}{4}$ ⊜ $\frac{6}{8}$.

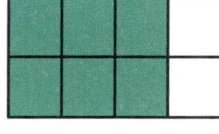

$\frac{3}{4} = \frac{6}{8}$ 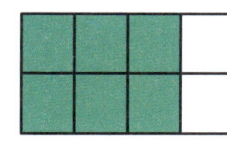 $\frac{6}{8}$

Another Way: Use a like numerator. Find an equivalent fraction for $\frac{6}{8}$ that has a numerator of 3.

Divide the numerator and the denominator of $\frac{6}{8}$ by __2__.

$$\frac{6}{8} = \frac{6 \div \boxed{2}}{8 \div \boxed{2}} = \frac{3}{4}$$

Compare $\frac{3}{4}$ and $\frac{3}{4}$.

$\frac{3}{4}$ ⊜ $\frac{3}{4}$

So, $\frac{6}{8}$ ⊜ $\frac{3}{4}$.

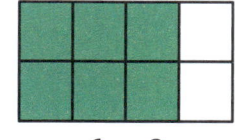

$\frac{6}{8} = \frac{3}{4}$ 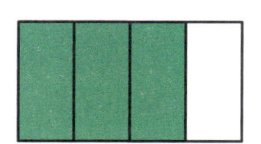 $\frac{3}{4}$

Compare. Use a model to help.

1. $\frac{3}{10}$ ◯ $\frac{1}{5}$

2. $\frac{4}{5}$ ◯ $\frac{2}{3}$

3. $\frac{5}{8}$ ◯ $\frac{2}{1}$

Chapter 7 | Lesson 5

Compare. Use a model to help.

4. $\frac{9}{10} \bigcirc \frac{97}{100}$

5. $\frac{3}{8} \bigcirc \frac{2}{6}$

6. $\frac{1}{3} \bigcirc \frac{4}{12}$

7. $\frac{7}{2} \bigcirc \frac{6}{5}$

8. $\frac{1}{10} \bigcirc \frac{2}{12}$

9. $\frac{3}{4} \bigcirc \frac{4}{6}$

10. **MP Structure** Compare $\frac{3}{8}$ and $\frac{1}{4}$ two different ways.

11. **Modeling Real Life** A sailor is making a ship in a bottle. The last thing he needs to do is seal the bottle with a cork stopper. He tries a $\frac{3}{4}$-inch cork stopper, but it is too small. Should he try a $\frac{1}{2}$-inch cork stopper or a $\frac{4}{5}$-inch cork stopper next? Explain.

12. **DIG DEEPER!** Order the lengths of hair donated from greatest to least.

Student	Hair Lengths Donated (feet)
Student A	$\frac{3}{4}$
Student B	$\frac{11}{12}$
Student C	$\frac{5}{6}$

Review & Refresh

13. Extend the pattern of shapes by repeating the rule "triangle, pentagon, octagon." What is the 48th shape in the pattern?

334

Name _____

Performance Task 7

1. a. Your art teacher wants you to complete the design below. Half of the squares are colored black. Complete the table. Then use the table to finish the design.

Color	Number of Squares	Fraction of Total Squares
Black		$\frac{1}{2}$
Red	10	
Orange		$\frac{1}{20}$
Yellow		$\frac{2}{25}$
Green	7	
Blue	8	
White		$\frac{3}{25}$

b. Which two colors cover the same portion of the design? Explain.

_____

2. Your teacher displays 30 designs in a rectangular array on the wall. Show two different ways your teacher can arrange the designs.

Chapter 7   335

Fraction Boss

Directions:
1. Divide the Fraction Boss Cards equally between both players.
2. Each player flips a Fraction Boss Card.
3. Players compare their fractions. The player with the greater fraction takes both cards.
4. If the fractions are equal, each player flips another card. Players compare their fractions. The player with the greater fraction takes all four cards.
5. The player with the most cards at the end of the round wins!

Player A	Player B

Name _____

Chapter Practice 7

7.1 Model Equivalent Fractions

1. Use the model to find an equivalent fraction for $\frac{3}{4}$.

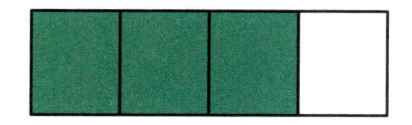

2. Use the number line to find an equivalent fraction for $\frac{2}{5}$.

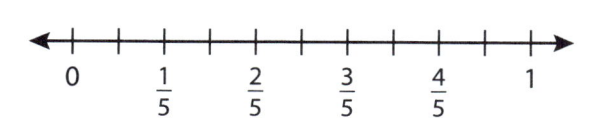

3. **Open-Ended** Write two equivalent fractions to describe the portion of the apples that are red.

7.2 Generate Equivalent Fractions by Multiplying

Find the equivalent fraction.

4. $\frac{3}{4} = \frac{\square}{12}$

5. $\frac{1}{2} = \frac{5}{\square}$

6. $\frac{8}{5} = \frac{\square}{100}$

Find an equivalent fraction.

7. $\frac{1}{6}$

8. $\frac{5}{5}$

9. $\frac{1}{4}$

Find two equivalent fractions.

10. $\frac{3}{2}$

11. $\frac{1}{3}$

12. $\frac{4}{5}$

Chapter 7   337

7.3 Generate Equivalent Fractions by Dividing

Find the equivalent fraction.

13. $\dfrac{3}{12} = \dfrac{\boxed{}}{4}$

14. $\dfrac{18}{100} = \dfrac{9}{\boxed{}}$

15. $\dfrac{20}{10} = \dfrac{\boxed{}}{2}$

Find an equivalent fraction.

16. $\dfrac{4}{6}$

17. $\dfrac{16}{4}$

18. $\dfrac{20}{8}$

Find two equivalent fractions.

19. $\dfrac{80}{100}$

20. $\dfrac{6}{12}$

21. $\dfrac{40}{4}$

22. **Modeling Real Life** You have 90¢. What fraction of a dollar, in tenths, do you have?

7.4 Compare Fractions Using Benchmarks

Compare. Use a model to help.

23. $\dfrac{7}{8} \bigcirc \dfrac{2}{5}$

24. $\dfrac{6}{10} \bigcirc \dfrac{4}{3}$

25. $\dfrac{1}{6} \bigcirc \dfrac{2}{12}$

7.5 Compare Fractions

Compare. Use a model to help.

26. $\dfrac{1}{4} \bigcirc \dfrac{3}{12}$

27. $\dfrac{2}{3} \bigcirc \dfrac{6}{10}$

28. $\dfrac{7}{8} \bigcirc \dfrac{5}{6}$

Cumulative Practice 1-7

1. Which number is a common factor of 12, 16, and 40?

   (A) 5
   (B) 8
   (C) 3
   (D) 4

2. Which statements describe the difference of 77,986 and 21,403?

   ☐ The difference is greater than 60,000.
   ☐ The difference is about 60,000.
   ☐ The difference is less than 60,000.
   ☐ The difference is 56,583.

3. Which product is between 5,050 and 5,100?

   (A) 652×8
   (B) 566×9
   (C) $1,023 \times 5$
   (D) $1,257 \times 4$

4. Use the advertisement to answer the question.

CAR FOR SALE
- ✓ 84,700 miles (rounded to the nearest 100)
- ✓ New tires
- ✓ Clean interior

What is the greatest number of miles that the car could have been driven?

   (A) 84,699
   (B) 84,750
   (C) 84,749
   (D) 84,650

Chapter 7

5. Which multiplication expressions could be represented by the area model?

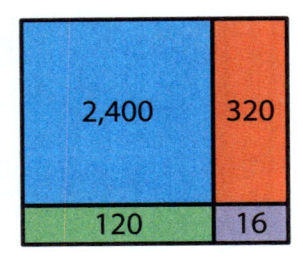

2,400	320
120	16

☐ 84 × 34

☐ 42 × 68

☐ 240 × 10

☐ 20 × 6

6. Your neighbor buys yogurt in packages of 4. If your neighbor only buys complete packages, how many yogurts could he have bought?

Ⓐ 2

Ⓑ 20

Ⓒ 34

Ⓓ 18

7. Which fraction makes the statement true?

$$? < \frac{6}{8}$$

Ⓐ $\frac{1}{4}$

Ⓑ $\frac{7}{8}$

Ⓒ $\frac{6}{6}$

Ⓓ $\frac{3}{4}$

8. Which division equation is represented by the counters?

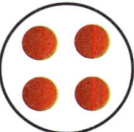

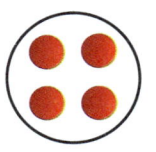

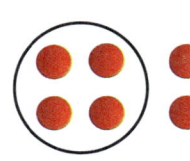

Ⓐ $6 \div 4 \stackrel{?}{=} 2$

Ⓑ $26 \div 6 \stackrel{?}{=} 2 \text{ R}4$

Ⓒ $26 \div 6 \stackrel{?}{=} 4$

Ⓓ $26 \div 6 \stackrel{?}{=} 4 \text{ R}2$

9. Which statements are true?

- [ ] All prime numbers are odd.
- [ ] A composite number cannot have 3 factors.
- [ ] 99 is a prime number.
- [ ] A prime number's factors are 1 and itself.
- [ ] 27 is a composite number.
- [ ] All even numbers greater than 1 are composite.

10. What is the quotient of 3,258 and 3?

11. Which model shows an equivalent fraction for the fraction shown by the model below?

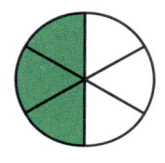

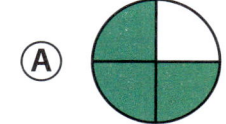

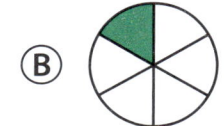

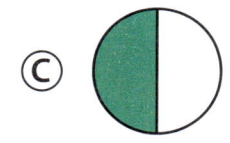

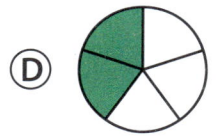

Ⓐ   Ⓑ

Ⓒ   Ⓓ

12. What is the product of 68 and 45?

Ⓐ 8,420

Ⓑ 3,060

Ⓒ 612

Ⓓ 27,360

13. Which pattern uses the same rule as the pattern below?

2, 10, 18, 26, 34, 42

Ⓐ 15, 23, 31, 39, 47, 55

Ⓑ 5, 25, 125, 625, 3,125, 15,625

Ⓒ 70, 62, 54, 46, 38, 30

Ⓓ 25, 34, 43, 52, 61, 70

Chapter 7

14. Which statements are true?

    ☐ $\frac{1}{5} \stackrel{?}{=} \frac{2}{10}$ ☐ $\frac{1}{3} \stackrel{?}{=} \frac{3}{9}$ ☐ $\frac{1}{4} \stackrel{?}{=} \frac{1}{8}$

    ☐ $\frac{2}{8} \stackrel{?}{=} \frac{1}{4}$ ☐ $\frac{4}{6} \stackrel{?}{=} \frac{2}{4}$ ☐ $\frac{6}{12} \stackrel{?}{=} \frac{1}{2}$

15. **Think Solve Explain**

    Part A At a summer camp, 67 students are in a line to rent kayaks. Each kayak can hold 4 people. How many kayaks will be full?

    Part B How many kayaks will be used?

    Part C How many students will be in the last kayak? Explain.

16. How many zeros will the product of 80 and 50 have?

    Ⓐ 1   Ⓑ 2

    Ⓒ 3   Ⓓ 4

17. There are 57 electronic books checked out of a library. There are 8 times as many printed books checked out as electronic books. How many total books are checked out of the library?

    Ⓐ 513   Ⓑ 4,113

    Ⓒ 122   Ⓓ 456

18. Which number is equal to 100,000 + 5,000 + 80 + 4?

    Ⓐ 1,584   Ⓑ 105,084

    Ⓒ 15,084   Ⓓ 105,840

Name _____

STEAM Performance Task 1-7

1. Sea level is the average level of the oceans on Earth. The global sea level is rising about $\frac{1}{8}$ inch each year.

   a. If this pattern continues, about how much will the sea level rise in 80 years?

   b. You read from another source that the sea level is rising about $\frac{1}{2}$ inch every 4 years. Did your source use the same fact as above? Explain.

   c. In 10 years, will the sea level rise more or less than an inch? Explain.

   Remember, there are 12 inches in one foot.

   d. How many years will it take the sea level to rise about 1 foot?

   e. Use the Internet or some other resource to learn about rising sea levels. Write one interesting fact that you learn.

Chapter 7   343

2. The gravitational pull of the moon affects the high and low tides of the oceans on Earth. Cities along the coasts use tide tables each day. Use the tide table to answer the questions.

   a. When swimming, why is it important to understand tide tables?

| Tide Table ||
Time of Day	Height of Water
7:00 A.M.	48 in.
10:00 A.M.	28 in.
1:00 P.M.	8 in.
4:00 P.M.	32 in.
7:00 P.M.	56 in.
10:00 P.M.	32 in.
1:00 A.M.	4 in.

   b. What are the common factors of the water heights?

   c. Make a picture graph of the water heights.

   Each ◯ = _____ inches.

   d. What pattern do you notice about the water heights? Explain.

344

Glossary

A

acute angle   [ángulo agudo]

An angle that is open less than a right angle

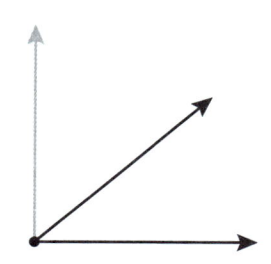

acute triangle   [triángulo acutángulo]

A triangle that has three acute angles

adjacent angles   [ángulos adyacentes]

Two angles that share a common side and a common vertex, but have no other points in common

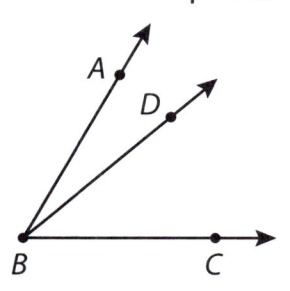

∠ABD and ∠DBC are adjacent angles.

angle   [ángulo]

Two rays or line segments that have a common endpoint

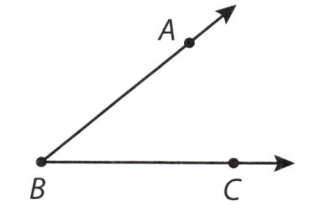

Label: ∠ABC, ∠CBA, ∠B

area   [área]

The amount of surface a figure covers

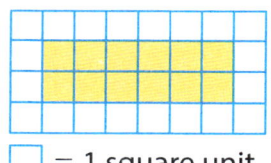

☐ = 1 square unit

The area of the rectangle is 12 square units.

B

benchmark   [punto de referencia]

A commonly used number that you can use to compare other numbers

Examples: $\frac{1}{2}$, 1

A1

C

common factor [factor común]

A factor that is shared by two or more given numbers

Factors of 8: ①, ②, ④, 8

common factors

Factors of 12: ①, ②, 3, ④, 6, 12

compatible numbers [números compatibles]

Numbers that are easy to multiply and are close to the actual numbers

24 × 31
↓   ↓
25 × 30

complementary angles [ángulos complementarios]

Two angles whose measures have a sum of 90°

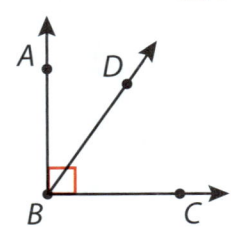

∠ABD and ∠DBC are complementary angles.

composite number [número compuesto]

A whole number greater than 1 with more than two factors

27

The factors of 27 are 1, 3, 9, and 27.

cup (c) [taza (tz)]

A customary unit used to measure capacity

The capacity of the measuring cup is 1 cup.

D

decimal [decimal]

A number with one or more digits to the right of the decimal point

0.3

0.04

0.59

decimal fraction [fracción decimal]

A fraction with a denominator of 10 or 100

$\frac{26}{100}$

$\frac{9}{10}$

$\frac{60}{100}$

decimal point [punto decimal]

A symbol used to separate the ones place and the tenths place in numbers, and to separate the whole dollars and the cents in money

0.1 $5.06

decimal point

degree (°) [grado (°)]

The unit used to measure angles

$1° = \frac{1}{360}$ of a circle

Distributive Property [propiedad distributiva]

$3 \times (5 + 2) = (3 \times 5) + (3 \times 2)$

$3 \times (5 - 2) = (3 \times 5) - (3 \times 2)$

divisible [divisible]

A number is divisible by another number when the quotient is a whole number and the remainder is 0.

$48 \div 4 = 12$ R0

So, 48 is divisible by 4.

E

endpoints [puntos extremos]

Points that represent the ends of a line segment or ray

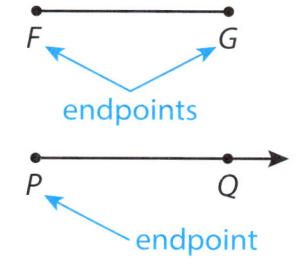

F — G
endpoints

P — Q →
endpoint

equiangular triangle [triángulo equiángulo]

A triangle that has three angles with the same measure

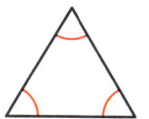

equilateral triangle [triángulo equilátero]

A triangle that has three sides with the same length

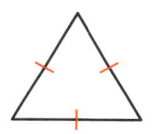

equivalent [equivalente]

Having the same value

$\frac{8}{8} = 1$

$3 = \frac{3}{1}$

$2 = \frac{4}{2} = \frac{6}{3}$

equivalent decimals [decimales equivalente]

Two or more decimals that have the same value

$0.40 = 0.4$

A3

equivalent fractions
[fracciones equivalentes]

Two or more fractions that name the same part of a whole

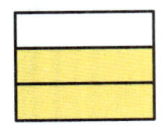

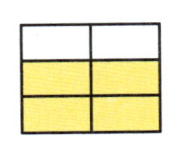

$$\frac{2}{3} = \frac{4}{6}$$

estimate **[estimación]**

A number that is close to an exact number

8,195 + 9,726 = ?

Exact Sum: 17,921   Estimate: 18,000

F

factor pair **[par de factores]**

Two factors that, when multiplied, result in a given product

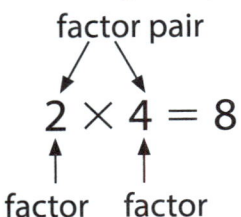

factor pair
$$2 \times 4 = 8$$
factor   factor

2 and 4 are a factor pair for 8.

formula **[fórmula]**

An equation that uses letters and numbers to show how quantities are related

$$P = (2 \times \ell) + (2 \times w)$$
$$A = \ell \times w$$

G

gallon (gal) **[galón (gal)]**

A customary unit used to measure capacity
There are 4 quarts in 1 gallon.

The capacity of the jug is 1 gallon.

H

hundredth **[centésimo]**

1 of 100 equal parts of a whole

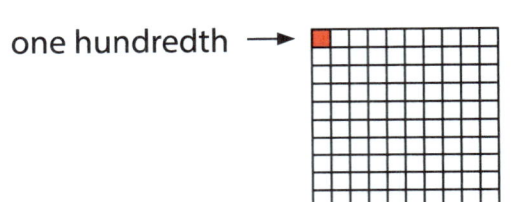

one hundredth →

hundredths place
[posición de los centésimos]

The second place to the right of the decimal point

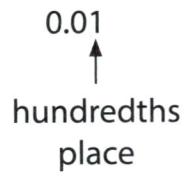

0.01
↑
hundredths place

I

intersecting lines [líneas secantes]

Lines that cross at exactly one point

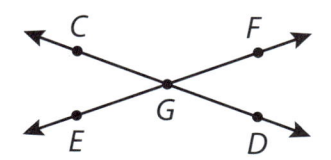

isosceles triangle [triángulo isósceles]

A triangle that has two sides with the same length

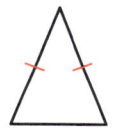

K

kilometer (km) [kilómetro (km)]

A metric unit used to measure length
There are 1,000 meters in 1 kilometer.

1 kilometer is about the length of 10 football fields including the end zones.

L

line [línea]

A straight path of points that goes on without end in both directions

Label: $\overleftrightarrow{CD}, \overleftrightarrow{DC}$

line of symmetry [línea de simetría]

A fold line that divides a shape into two parts that match exactly

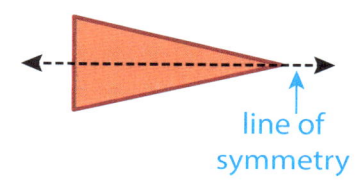

line of symmetry

line segment [segmento lineal]

A part of a line that includes two endpoints and all of the points between them

Label: $\overline{FG}, \overline{GF}$

line symmetry [simetría lineal]

The symmetry that a shape has when it can be folded on a line so that two parts match exactly

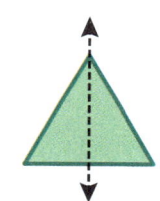

A5

M

mile (mi) [milla (mi)]

A customary unit used to measure length
There are 1,760 yards in 1 mile.

When walking briskly, you can walk 1 mile in about 20 minutes.

millimeter (mm) [milímetro (mm)]

A metric unit used to measure length

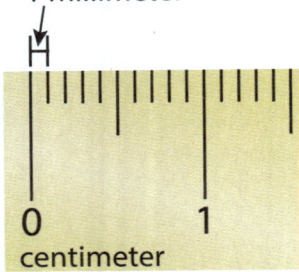

There are 10 millimeters in 1 centimeter.

mixed number [número mixto]

Represents the sum of a whole number and a fraction less than 1

Examples: $2\frac{1}{3}, 1\frac{4}{5}, 5\frac{3}{10}$

multiple [múltiplo]

The product of a number and any other counting number

$1 \times 4 = 4$
$2 \times 4 = 8$
$3 \times 4 = 12$
$4 \times 4 = 16$

↑ multiples of 4

O

obtuse angle [ángulo obtuso]

An angle that is open more than a right angle and less than a straight angle

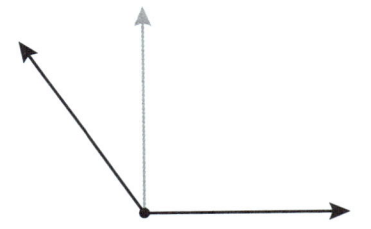

obtuse triangle [triángulo obtusángulo]

A triangle that has one obtuse angle

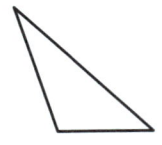

ones period [período de las unidades]

The first period in a number

Thousands Period			Ones Period		
Hundreds	Tens	Ones	Hundreds	Tens	Ones
8	1	5,	7	9	6

ounce (oz) [onza (oz)]

A customary unit used to measure weight

A slice of bread weighs about 1 ounce.

A6

P

parallel lines [líneas paralelas]

Lines that never intersect

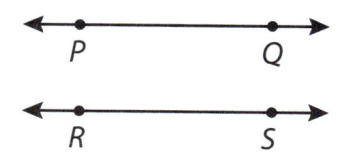

Label: $\overleftrightarrow{PQ} \parallel \overleftrightarrow{RS}$

parallelogram [paralelogramo]

A quadrilateral that has two pairs of parallel sides

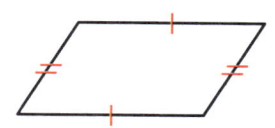

partial products [productos parciales]

The products found by breaking apart a factor into ones, tens, hundreds, and so on, and multiplying each of these by the other factor

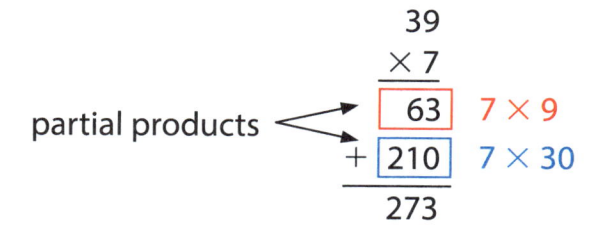

$$\begin{array}{r} 39 \\ \times 7 \\ \hline 63 \\ + 210 \\ \hline 273 \end{array}$$

partial products: $63 = 7 \times 9$, $210 = 7 \times 30$

partial quotients [cocientes parciales]

A division strategy in which quotients are found in parts until the remainder is less than the divisor

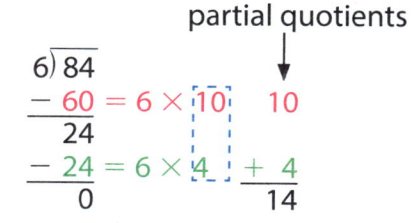

$$\begin{array}{r} 6\overline{)84} \\ -60 = 6 \times 10 \quad 10 \\ \hline 24 \\ -24 = 6 \times 4 \quad + 4 \\ \hline 0 \quad\quad\quad 14 \end{array}$$

partial quotients

perimeter [perímetro]

The distance around a figure

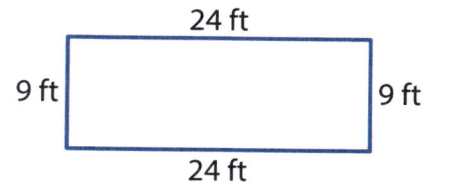

24 ft, 9 ft, 9 ft, 24 ft

The perimeter is 66 feet.

period [período]

Each group of three digits separated by commas in a multi-digit number

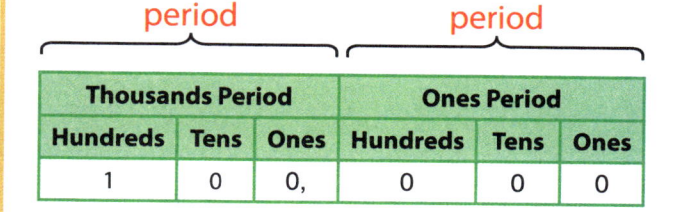

Thousands Period			Ones Period		
Hundreds	Tens	Ones	Hundreds	Tens	Ones
1	0	0,	0	0	0

perpendicular lines [líneas perpendiculares]

Lines that intersect to form four right angles

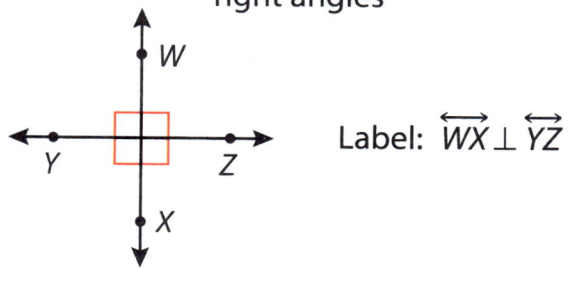

Label: $\overleftrightarrow{WX} \perp \overleftrightarrow{YZ}$

pint (pt) [pinta (pt)]

A customary unit used to measure capacity
There are 2 cups in 1 pint.

The capacity of the carton is 1 pint.

A7

place value chart [gráfico de valor posicional]

A chart that shows the value of each digit in a number

Thousands Period			Ones Period		
Hundreds	Tens	Ones	Hundreds	Tens	Ones
2	8	5,	7	4	3

point [punto]

An exact location in space

A

Label: point A

pound (lb) [libra (lb)]

A customary unit used to measure weight
There are 16 ounces in 1 pound.

A loaf of bread weighs about 1 pound.

prime number [número primo]

A number greater than 1 with exactly two factors, 1 and itself

11
The factors of 11 are 1 and 11.

protractor [transportador]

A tool for measuring and drawing angles

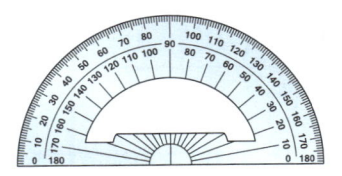

Q

quart (qt) [cuarto de galón (qt)]

A customary unit used to measure capacity
There are 2 pints in 1 quart.

The capacity of the carton is 1 quart.

R

ray [semirrecta]

A part of a line that has one endpoint and goes on without end in one direction

P   Q

Label: $\overrightarrow{PQ}$

rectangle [rectángulo]

A parallelogram that has four right angles

A8

remainder [resto]

The amount left over when a number cannot be divided evenly

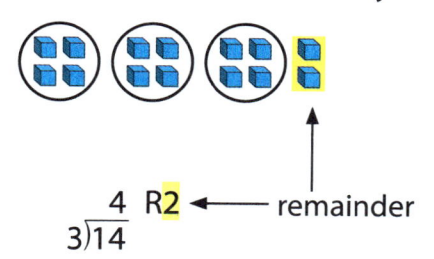

$$3\overline{)14} \quad \begin{array}{r} 4 \ \ R2 \end{array} \leftarrow \text{remainder}$$

---

rhombus [rombo]

A parallelogram that has four sides with the same length

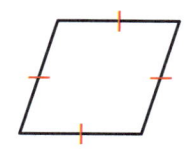

right angle [ángulo recto]

An L-shaped angle

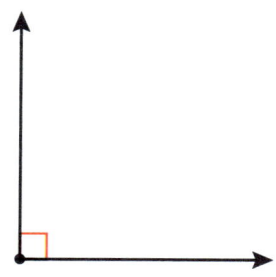

---

right triangle
[triángulo rectángulo]

A triangle that has one right angle

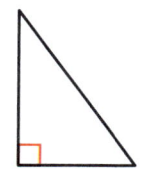

---

rule [regla]

Tells how numbers or shapes in a pattern are related

Rule: Add 3.
3, 6, 9, 12, 15, 18, 21, 24, . . .

Rule: triangle, hexagon, square, rhombus

S

scalene triangle [triángulo escaleno]

A triangle that has no sides with the same length

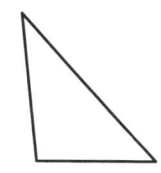

second (sec) [segundo (seg)]

A unit of time

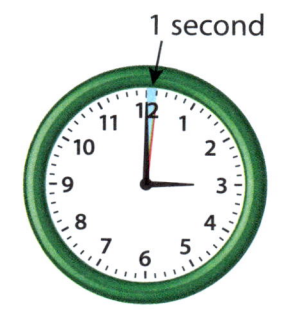

1 second

There are 60 seconds in 1 minute.

---

square [cuadrado]

A parallelogram that has four right angles and four sides with the same length

A9

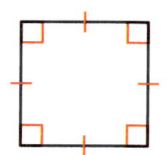

straight angle   [ángulo llano]

An angle that forms a straight line

---

supplementary angles
[ángulos suplementarios]

Two angles whose measures have a sum of 180°

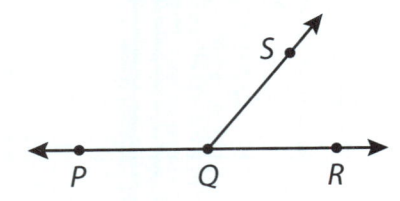

∠PQS and ∠SQR are supplementary angles.

T

tenth   [décimo]

1 of 10 equal parts of a whole

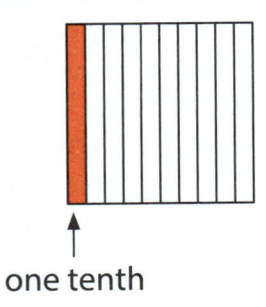

one tenth

---

tenths place
[posición de los décimos]

The first place to the right of the decimal point

0.1
↑
tenths place

---

thousands period
[período de las milésimas]

The period after the ones period in a number

Thousands Period			Ones Period		
Hundreds	Tens	Ones	Hundreds	Tens	Ones
8	1	5,	7	9	6

---

ton (T)   [tonelada (T)]

A customary unit used to measure weight There are 2,000 pounds in 1 ton.

A small compact car weighs about 1 ton.

---

trapezoid   [trapecio]

A quadrilateral that has exactly one pair of parallel sides

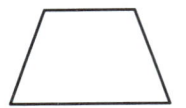

U

unit fraction   [fracción unitaria]

Represents one equal part of a whole

Examples:

$\frac{1}{2}$ $\frac{1}{5}$

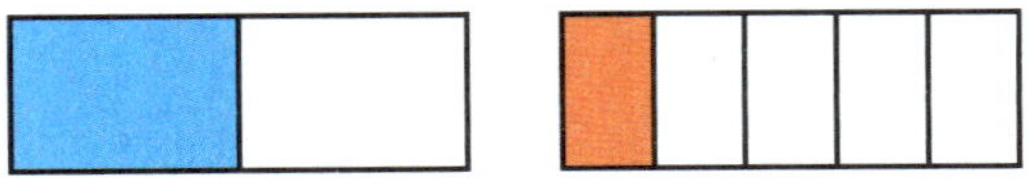

A10

V

vertex [vértice]

The endpoint at which two rays or line segments of an angle meet

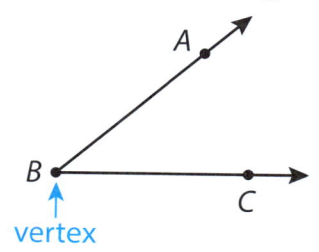

↑
vertex

Index

A

Acute angles
   classifying triangles by, 667–672
   definition of, 600
   finding unknown measures of, 635–640
   identifying, drawing, and naming, 600–604
   measuring
      using degrees, 612–616
      using pattern blocks, 617–622

Acute triangles
   classifying, 668–672
   definition of, 668

Addition
   of angle measures, 629–634
   Associative Property of, 384
   Commutative Property of, 384
   comparing numbers using, 69–74
   using compensation, 51, 52
   using "count on" strategy, 51, 53, 55
   of decimal fractions and decimals, 469–474
   estimating sums in, 33–38
      to check reasonableness of answer, 39–44
   of fractions
      using area models, 347–352
      with like denominator, 359–364
      in mixed numbers, 383–388
      in multiples of fractions, 415–420
      in multiples of unit fractions, 409–414
      using number line, 348, 351
      sums of, writing fractions as, 353–358, 409–420
   of mixed measures, 543–548
   of money amounts, 481–486
   of multi-digit numbers, 39–44
      strategies for, 51–56
   using number line, 53, 55
   using partial sums, 51, 52
   patterns in, 286–290
   relationship between subtraction and, 390
   two-step word problems solved using, 57–62

Adjacent angles
   adding measures of, 630–634
   definition of, 630
   finding unknown measures with, 636–640

Angles
   acute (*See* Acute angles)
   adding measures of, 629–634
   adjacent (*See* Adjacent angles)
   classifying quadrilaterals by, 673–678
   classifying triangles by, 667–672
   complementary (*See* Complementary angles)
   definition of, 600
   finding unknown measures of, 635–640
   identifying, drawing, and naming, 599–604
   measuring
      using degrees, 611–616
      using fractional parts of circle, 612–616
      using pattern blocks, 617–622
   measuring and drawing with protractor, 623–628
   obtuse (*See* Obtuse angles)
   one-degree, 612
   right (*See* Right angles)
   straight (*See* Straight angles)
   supplementary (*See* Supplementary angles)

Another Way, *Throughout. For example, see:* 34, 52, 144, 152, 162, 218, 274, 306, 354, 422

Apply and Grow: Practice, *In every lesson. For example, see:* 5, 35, 71, 145, 201, 263, 307, 349, 411, 447

A13

Area(s)
    definition of, 570
    of rectangles
        finding unknown measures from, 575–580
        formula for, 569–574
        solving word problems involving, 581–586

Area models
    adding fractions using, 347–352
    division using, 217–228
    finding factor pairs in, 261–266
    multiplication using
        Distributive Property and, 87–98, 161–166
        expanded form and, 93–98
        place value and, 99
        with two-digit numbers, 155–166, 179, 180
    subtracting fractions using, 365–370

Associative Property of Addition, 384

Associative Property of Multiplication, 117–122
    definition of, 118
    multiples of fractions and, 416, 418
    multiplying by tens using, 143–148
    multiplying fractions by whole numbers using, 422, 425
    multiplying two-digit numbers using, 179, 183

B

Balance scale, for measuring mass, 501

Benchmarks
    comparing fractions using, 323–328
    definition of, 324

Breaking apart factors
    in Distributive Property, 87–92
    in finding partial products, 99–104, 155–160

Breaking apart fractions, 353–358

C

Capacity
    in customary units, 519–524
    in metric units, 501–506
    in mixed measures, 544–548

Centimeters
    definition of, 496
    equivalent length in, 495–500
    finding area in, 570–573
    finding perimeter in, 564–568

Challenge, *See* Dig Deeper

Chapter Practice, *In every chapter. For example, see:* 29–30, 65–66, 131–134, 193–196, 255–258, 299–302, 337–338, 403–406, 441–442, 489–492

Charts
    hundred, identifying prime numbers on, 281
    place value (*See* Place value charts)

Check for Reasonableness, *Throughout. For example, see:* 39, 57–62, 81–86, 168–169, 231, 429, 509, 582

Circles
    adding angle measures within, 629–634
    fractional parts of, measuring angles using, 612–616
    line symmetry/lines of symmetry in, 651–653

Classifying shapes
    quadrilaterals, 673–678
    triangles
        by angles, 667–672
        by sides, 661–666, 668

Clock, measuring elapsed time on, 537–542

Coins
    amounts in fractions and decimals, 475–480
    operations with, 481–486

Comma, separating periods in numbers, 4

Common Errors, *Throughout. For example, see:* 45, T-79, 117, T-175, T-294, T-372, T-422, T-504, T-565, T-596

Common factors
  definition of, 318
  dividing by, to find equivalent fractions, 317–322

Common Misconceptions, *Throughout. For example, see:* T-22, 39, T-268, 279, T-303C, T-330, T-458

Commutative Property of Addition, 384

Commutative Property of Multiplication
  using, 117–122
  comparisons using, 70
  definition of, 118

Comparing decimals, 463–468
  on number line, 464, 465, 467
  on place value chart, 464, 466, 467

Comparing fractions
  using benchmarks, 323–328
  using equivalent fractions, 329–334
  on number line, 324, 327

Comparing numbers
  using addition, 69–74
  multi-digit, 15–20
  using multiplication, 69–74
  using place value, 3–8

Comparison sentences, 69–74

Compatible numbers
  definition of, 150, 206
  estimating products using, 149–154
  estimating quotients using, 205–210

Compensation
  addition using, 51, 52
  subtraction using, 51, 52, 54, 55

Complementary angles
  definition of, 636
  finding measures of, 636–640

Composite numbers
  definition of, 280
  identifying, 279–284

"Count on" strategy
  addition using, 51, 53, 55
  subtraction using, 51, 52

Cross-Curricular Connections, *In every lesson. For example, see:* T-7, T-49, T-109, T-171, T-295, T-309, T-399, T-603

Cumulative Practice, 135–138, 339–342, 555–558, 683–686

Cups (c)
  definition of, 520
  equivalent capacity in, 519–524
  in mixed measures, 544, 545

Customary units
  of capacity
    definitions of, 520
    equivalence of, 519–524
  of length
    definition of, 508
    equivalence of, 507–512
  of weight
    definitions of, 514
    equivalence of, 513–518

D

Data, line plots of, 525–530

Days
  definition of, 532
  equivalent time in, 532–536
  in mixed measures, 545, 547, 548

Decimal(s)
  adding, 469–474
  comparing, 463–468
    on number line, 464, 465, 467
    on place value chart, 464, 466, 467
  definition of, 446
  equivalent
    comparing, 464
    definition of, 458
  hundredths
    definition of, 452
    extending place value chart to include, 452, 455
    money amounts in, 475–480
    plotting on number line, 457, 459, 461

A15

understanding, 451–456
writing as equivalent fractions and decimals, 457–462
money written in, 475–480
plotting on number line, 457, 459, 461
tenths
definition of, 446
extending place value chart to include, 446, 449
money amounts in, 475–480
plotting on number line, 457
understanding, 445–450
writing as equivalent fractions and decimals, 457–462
writing in expanded form, 459

Decimal form, writing in, 457–462

Decimal fractions
adding, 469–474
definition of, 452

Decimal point
amount of money with, 475–480
definition of, 446

Decomposition
of angles
adding measures in, 629–634
finding unknown measures in, 635–640
of factors
in Distributive Property, 87–92
in finding partial products, 99–104, 155–160
of fractions, 353–358

Define It, In every chapter. For example, see: 2, 32, 68, 142, 198, 260, 304, 346, 408, 444

Degrees
definition of, 612
measuring angles using, 611–616
relationship between fractional parts of circle and, 612–616
symbol for, 612

Denominators
in comparing fractions, 329–334
in decimal fractions, 452

in generating equivalent fractions
by dividing, 317–322
by multiplying, 311–316
like
adding fractions with, 359–364
adding mixed numbers with, 383–388
subtracting fractions with, 371–376
subtracting mixed numbers with, 389–394
writing fractions as multiple of unit fractions with, 409–414

Differences, estimating, *See also* Subtraction
with multi-digit numbers, 45–50
using rounding, 33–38

Differentiation, *see* Scaffolding Instruction

Dig Deeper, *Throughout. For example, see:* 8, 35, 148, 201, 264, 308, 350, 411, 448, 497

Dimes, as tenths, 445, 476

Distances, line plot of, 528

Distributive Property, 87–92, 117–122
area models with, 87–98, 161–166
definition of, 88, 118
expanded form and, 93–98
finding perimeter with, 568
multiplying two-digit numbers with, 161–166, 179, 183
multiplying whole numbers and mixed numbers using, 427–432

Divisibility
definition of, 268
finding factor pairs in
using division, 267–272
using models, 261–266
of prime and composite numbers, 279–284

Divisibility rules, 268, 274, 280, 283

Division
using area models, 217–228
of composite numbers, 279–284
estimating quotients in, 205–210
factor pairs in
definition of, 262
using division to find, 267–272

A16

using models to find, 261–266
generating equivalent fractions using, 317–322
using models to find quotients and remainders in, 211–216
of money amounts, 481–486
of multi-digit numbers, by one-digit numbers, 235–240
by one-digit numbers, 241–246
multi-digit numbers, 235–240
tens, hundreds, or thousands, 199–204
two-digit numbers, 229–234
using partial quotients, 217–222
with remainders, 223–228
patterns in, 286–290
using place value
in dividing by one-digit numbers, 241–246
multi-digit numbers by one-digit numbers, 235–240
tens, hundreds, and thousands, 199–204
two-digit numbers by one-digit numbers, 229–234
of prime numbers, 279–284
of two-digit numbers, by one-digit numbers, 229–234
word problems solved using, 247–252

Division facts
to divide by tens, hundreds, or thousands, 200
to estimate quotients, 205–210

Dollar(s)
amounts in fractions and decimals, 475–480
operations with, 481–486

Dollar sign, 476

Drawing
angles, 599–604, 623–628
intersecting, parallel, and perpendicular lines, 605–610
points, lines, and rays, 593–598
symmetric shapes, 655–660

E

Elapsed time
measuring on number line, 538, 540
solving word problems involving, 537–542

ELL Support, *In every lesson. For example, see:* T-2, T-87, T-173, T-260, T-323, T-445, T-587, T-679

Endpoints
definition of, 594
vertices as, 600

Equal groups, remainders in division into, 211–216

Equiangular triangles
classifying, 668–672
definition of, 668

Equilateral triangles
classifying, 662–666, 668
definition of, 662

Equivalent, definition of, 306

Equivalent decimals
comparing, 464
definition of, 458

Equivalent fractions
adding decimal fractions and decimals using, 469–474
adding mixed numbers using, 383–388
comparing fractions using, 329–334
definition of, 306
generating
by dividing, 317–322
by multiplying, 311–316
modeling and writing, 305–310
on number line, 306–310, 312, 319, 321
writing tenths and hundredths as, 457–462

Equivalent measurements
capacity
in customary units, 519–524
in metric units, 501–506

A17

length
  in customary units, 507–512
  in metric units, 495–500
mass, 501–506
mixed measures, 543–548
time, 531–536
weight, 513–518
Error Analysis, *See* You Be the Teacher
Estimate, definition of, 34, 82
Estimating differences
  using rounding, 33–38
  in subtracting multi-digit numbers, 45–50
Estimating products, 149–154
  choosing expression for, 149
  using compatible numbers, 149–154
  using rounding, 81–86, 149–154
Estimating quotients, 205–210
Estimating sums
  to check reasonableness of answer, 39–44
  using rounding, 33–38
Expanded form
  fractions and decimals in, 459
  multi-digit numbers in, 10–14
  multiplication using, 93–98
Explain, *Throughout. For example, see:* 21, 58, 160, 263, 364, 426, 549, 586, 635, 682
Explore and Grow, *In every lesson. For example, see:* 3, 33, 69, 143, 199, 261, 305, 347, 409, 445

F

Factor(s), *See also* Multiplication
  breaking apart
    in Distributive Property, 87–92
    in finding partial products, 99–104, 155–160
  common
    definition of, 318
    dividing by, to find equivalent fractions, 317–322
  definition of, 274
  exactly two (prime number), 279–284
  more than two (composite number), 279–284
  relationship between multiples and, 273–278
  understanding, 261–266
Factor pairs
  definition of, 262
  using division to find, 267–272
  using models to find, 261–266
Feet (foot)
  definition of, 508
  equivalent length in, 507–512
  finding area in, 570–574
  finding perimeter in, 565–568
  in mixed measures, 543–548
Fewer, how many (comparison), 69–74
Folding, line symmetry in, 649–654
Formative Assessment, *Throughout. For example, see:* T-6, T-114, T-214, T-326, T-424, T-504
Formula
  for area of rectangle, 569–574
  definition of, 564
  for perimeter of rectangle, 563–568
Four-digit numbers, *See also* Multi-digit numbers
  multiplying by one-digit numbers, 111–116
Fraction(s)
  adding
    using area models, 347–352
    with like denominator, 359–364
    in mixed numbers, 383–388
    in multiples of fractions, 415–420
    in multiples of unit fractions, 409–414
    using number line, 348, 351
    sums of, writing fractions as, 353–358, 409–420
  comparing
    using benchmarks, 323–328
    using equivalent fractions, 329–334
    on number line, 324, 327
  decimal
    adding, 469–474

definition of, 452
and decimals
understanding hundredths, 451–456
understanding tenths, 445–450
writing in fraction and decimal forms, 457–462
decomposing, 353–358
equivalent (*See* Equivalent fractions)
money written in, 475–480
multiplying
fractions and whole numbers, 421–426
whole numbers and mixed numbers, 427–432
subtracting
using area models, 365–370
with like denominator, 371–376
in mixed numbers, 389–394
using number line, 366, 369, 372
unit (*See* Unit fractions)
word problems solved using, 395–400, 433–438
writing as mixed numbers/writing mixed numbers as, 377–382
writing in expanded form, 459
Fraction form, writing in, 457–462
Fraction strips, for comparing fractions, 324, 327

G

Gallons (gal)
definition of, 520
equivalent capacity in, 519–524
in mixed measures, 544, 545, 547
Games, *In every chapter. For example, see:* 28, 64, 130, 192, 254, 298, 336, 402, 440, 488
Grams (g)
definition of, 502
equivalent mass in, 501–506

H

Half of shape, for drawing symmetric shapes, 655–660
Height
line plot of, 526
mixed measures of, 543–548
Hexagon pattern blocks, measuring angles using, 617
Higher Order Thinking, *See* Dig Deeper
Homework & Practice, *In every lesson. For example, see:* 7–8, 37–38, 73–74, 147–148, 203–204, 265–266, 309–310, 351–352, 413–414, 449–450
Hours
definition of, 532
equivalent time in, 531–536
measuring elapsed time in, 537–542
in mixed measures, 544, 546, 547, 548
How many fewer, 69–74
How many more, 69–74
How many times, 69–74
Hundred chart, identifying prime numbers on, 281
Hundreds
dividing, 199–204
multiplying, using place value, 75–80
regrouping
in dividing by one-digit numbers, 241–246
in dividing multi-digit numbers by one-digit numbers, 236–240
in multiplying multi-digit numbers by one-digit numbers, 111–116
in multiplying two-digit numbers, 174, 177–180
rounding to nearest
estimating products by, 81–86
in estimating sums and differences, 33–38
using place value, 21–26

A19

Hundreds place, 4–8
    in comparing numbers, 15–20
    in reading and writing numbers, 10–14

Hundredths
    definition of, 452
    extending place value chart to include, 452, 455
    money amounts in, 475–480
    plotting on number line, 457, 459, 461
    understanding, 451–456
    writing as equivalent fractions and decimals, 457–462

Hundredths place
    comparing decimals to, 463–468
    definition of, 452

Hyphens, for writing numbers, 10

I

Inches
    definition of, 508
    equivalent length in, 507–512
    finding area in, 570–574
    finding perimeter in, 564–568
    in mixed measures, 543, 544, 546

Intersecting lines
    definition of, 606
    identifying and drawing, 605–610

Inverse operations, 200

Isosceles triangles
    classifying, 662–666
    definition of, 662

K

Kilograms (kg)
    definition of, 502
    equivalent mass in, 501–506

Kilometers (km)
    definition of, 496
    equivalent length in, 496–500

L

Learning Target, *In every lesson. For example, see:* 3, 33, 69, 143, 199, 261, 305, 347, 409, 445

Length
    in area formula, 569–574
    line plot of, 525, 527
    measuring
        in customary units, 507–512
        in metric units, 495–500
        in mixed measures, 544–548
    in perimeter formula, 563–568
    of sides of figure
        classifying quadrilaterals by, 673–678
        classifying triangles by, 661–666, 668
        finding unknown of rectangles, 575–580

Letters representing numbers
    in division problems, 247–252
    in multiplication problems, 123–128, 185–190

Like denominators
    adding fractions with, 359–364
    adding mixed numbers with, 383–388
    subtracting fractions with, 371–376
    subtracting mixed numbers with, 389–394
    writing fractions as multiple of unit fractions with, 409–414

Line(s)
    definition of, 594
    identifying, drawing, and naming, 593–598
    intersecting
        definition of, 606
        identifying and drawing, 605–610
    parallel
        definition of, 606
        identifying and drawing, 605–610
        symbol for, 606, 609
    perpendicular
        definition of, 606

identifying and drawing, 605–610
symbol for, 606, 609
Line of symmetry
definition of, 650
for drawing symmetric shapes, 655–660
identifying, 650–654
Line plots
making and interpreting, 525–530
scale of, 526
title of, 526
Line segments
definition of, 594
identifying, drawing, and naming, 593–598
Line symmetry
definition of, 650
identifying shapes with, 649–654, 676
Liters (L)
definition of, 502
equivalent capacity in, 501–506
Logic, *Throughout. For example, see:* 14, 65, 95, 107, 181, 278, 300, 414, 524, 595

M

Mass, in metric units, 501–506
Math Musicals, *In every chapter. For example, see:* 4, 40, 82, 268, 372, 520
Measurement equivalence
capacity
in customary units, 519–524
in metric units, 501–506
length
in customary units, 507–512
in metric units, 495–500
mass, 501–506
mixed measures, 543–548
time, 531–536
weight, 513–518
Meter stick, 495
Meters
definition of, 496
equivalent length in, 495–500

finding area in, 570, 571, 573
finding perimeter in, 565–568
Metric units
length
definitions of, 496
equivalence of, 495–500
mass and capacity
definitions of, 502
equivalence of, 501–506
prefixes of, 497
Miles (mi)
definition of, 508
equivalent length in, 508–512
in mixed measures, 545, 547
Milliliters (mL)
definition of, 502
equivalent capacity in, 501–506
Millimeters (mm)
definition of, 496
equivalent length in, 495–500
Minutes
definition of, 532
equivalent time in, 531–536
measuring elapsed time in, 537–542
in mixed measures, 544–548
Mixed measures, adding and subtracting, 543–548
Mixed numbers
adding, 383–388
decimals and
understanding hundredths, 451–456
understanding tenths, 445–450
definition of, 378
multiplying by whole numbers, 427–432
on number line, 378, 380, 382
subtracting, 389–394
word problems solved using, 395–400, 433–438
writing as fractions/writing fractions as, 377–382
Model(s)
for adding fractions, 347–352
area (*See* Area models)
for finding factor pairs, 261–266

A21

for finding quotients and remainders, 211–216
for subtracting fractions, 365–370

Modeling
of Distributive Property, 87–92
of equivalent fractions, 305–310
of fractions and mixed numbers, 377–382

Modeling Real Life, *In every lesson. For example, see:* 8, 38, 74, 148, 204, 266, 310, 352, 414, 450

Money
amounts in fractions and decimals, 475–480
operations with, 481–486

Months
definition of, 532
equivalent time in, 532–536
in mixed measures, 545

More, how many (comparison), 69–74

Multi-digit numbers
adding, 39–44
strategies for, 51–56
comparing, 15–20
dividing by one-digit numbers, 235–240
identifying values of digits in, 3–8
multiplying by one-digit numbers, 111–116
reading and writing, 9–14
rounding, using place value, 21–26
standard, word, and expanded forms of, 10–14
subtracting, 45–49
strategies for, 51–56

Multiple Representations, *Throughout. For example, see:* 4, 52, 144, 162, 180, 330, 356, 659

Multiples
of 10, multiplication by, 143–148
using number line, 144
using place value, 75–80, 143–148
using properties, 143–148
definition of, 274
relationship between factors and, 273–278

of unit fractions
on number line, 410, 413, 417
writing fractions as, 409–414
writing multiples of fractions as, 415–420

Multiplication
in area formula, 569–574
using area models
Distributive Property and, 87–98, 161–166
expanded form and, 93–98
place value and, 99
two-digit numbers, 155–166, 179, 180
Associative Property of, 117–122
definition of, 118
multiples of fractions and, 416, 418
multiplying by tens using, 143–148
multiplying fractions by whole numbers using, 422, 425
multiplying two-digit numbers using, 179, 183
Commutative Property of, 70, 117–122
comparing numbers using, 69–74
Distributive Property of, 87–92, 117–122
area models with, 87–98, 161–166
definition of, 88, 118
expanded form and, 93–98
finding perimeter with, 568
multiplying two-digit numbers with, 161–166, 179, 183
multiplying whole numbers and mixed numbers using, 427–432
estimating products in, 149–154
choosing expression for, 149
using compatible numbers, 149–154
using rounding, 81–86, 149–154
of fractions
fractions and whole numbers, 421–426
whole numbers and mixed numbers, 427–432
generating equivalent fractions using, 311–316

letters representing unknown numbers in, 123–128, 185–190
of money amounts, 481–486
of multi-digit numbers, by one-digit numbers, 111–116
by multiples of 10, 143–148
  using number line, 144
  using place value, 75–80, 143–148
  using properties, 143–148
on number line, by multiples of 10, 144
using partial products, 99–104
  with area models, 155–166, 179, 180
  multi-digit numbers by one-digit numbers, 111–116
  two-digit numbers, 167–172, 179, 180
  two-digit numbers by one-digit numbers, 105–110
patterns in, 286–290
in perimeter formula, 563–568
using place value
  by multiples of 10, 75–80, 143–148
  with partial products, 99–104
  by tens, hundreds, or thousands, 75–80
  two-digit numbers, 167–172, 179, 180
using properties, 117–122
of two-digit numbers, 173–178
  using area models, 155–166, 179, 180
  using Associative Property, 179, 183
  using Distributive Property, 161–166, 179, 183
  estimating products of, 149–154
  by multiples of 10, 143–148
  by one-digit numbers, 105–110
  using partial products, 167–172, 179, 180
  using place value, 167–172, 179, 180
  practicing strategies of, 179–184
  using regrouping, 174, 177–180
  word problems solved using, 185–190
word problems solved using
  comparisons, 69–74
  multi-step, 123–128, 185–190
  with two-digit numbers, 185–190

N

Naming
angles, 599–604
lines, line segments, and rays, 593–598

Number(s), *See specific types of numbers*

Number line
adding on, 53, 55
  decimal fractions and decimals, 469
  fractions, 348, 351
comparing decimals using, 464–465, 467
comparing fractions using, 324, 327
elapsed time on, 538, 540
finding equivalent fractions using, 306–310, 312, 319, 321
mixed numbers on, 378, 380, 382
multiples of unit fractions on, 410, 413, 417
multiplying on, by multiples of 10, 144, 148
plotting decimals and fractions on, 457, 459, 461
rounding number on, 22, 26
subtracting on, 52
  fractions, 366, 369, 372
weight on, 513

Number patterns
creating and describing, 285–290
definition of, 286

Number Sense, *Throughout. For example, see:* 8, 38, 71, 169, 204, 272, 328, 417, 453, 497

Numerators
adding, in fractions with like denominator, 359–364
in comparing fractions, 329–334
in generating equivalent fractions
  by dividing, 317–322
  by multiplying, 311–316
multiplying by whole numbers, 421–426
subtracting, in fractions with like denominator, 371–376

A23

O

Obtuse angles
    classifying triangles by, 667–672
    definition of, 600
    finding unknown measures of, 635–640
    identifying, drawing, and naming, 600–604
    measuring
        using degrees, 613–616
        using pattern blocks, 617–622

Obtuse triangles
    classifying, 668–672
    definition of, 668

One or ones (1)
    as factor, in prime numbers, 279–284
    in number patterns, 286, 289
    regrouping
        in dividing by one-digit numbers, 241–246
        in dividing multi-digit numbers by one-digit numbers, 236–240
        in dividing two-digit numbers by one-digit numbers, 230–234
        in multiplying multi-digit numbers by one-digit numbers, 111–116
        in multiplying two-digit numbers, 174, 177–180
        in multiplying two-digit numbers by one-digit numbers, 105–110
    whole dollar as, 476

One-degree angle, 612

One-digit numbers
    division by, 241–246
        multi-digit numbers, 235–240
        tens, hundreds, or thousands, 199–204
        two-digit numbers, 229–234
    multiplication by
        multi-digit numbers, 111–116
        two-digit numbers, 105–110

Ones period, 4–8
    in comparing numbers, 15–20
    definition of, 4
    in reading and writing numbers, 10–14

Ones place, 4–8
    in comparing numbers, 15–20
    in reading and writing numbers, 10–14

Open-Ended, *Throughout. For example, see:* 23, 74, 151, 290, 307, 349, 423, 465, 515, 574

Organize It, *In every chapter. For example, see:* 2, 32, 68, 142, 198, 260, 304, 346, 408, 444

Ounces (oz)
    definition of, 514
    equivalent weight in, 513–518
    in mixed measures, 545–547

P

Parallel lines
    definition of, 606
    identifying and drawing, 605–610
    symbol for, 606, 609

Parallel sides, of quadrilaterals, 674–678

Parallelograms
    classifying, 674–678
    definition of, 674

Partial products
    definition of, 100
    finding in any order, 100
    multiplication using, 99–104
        with area models, 155–166, 179, 180
        multi-digit numbers by one-digit numbers, 111–116
        two-digit numbers, 167–172, 179, 180
        two-digit numbers by one-digit numbers, 105–110

Partial quotients
    definition of, 218
    division using, 217–222
        with remainders, 223–228

Partial sums, addition using, 51, 52

Pattern blocks, measuring angles using, 617–622

Patterns
number, 285–290
shape, 291–296
Patterns, *Throughout. For example, see:* 76, 80, 184, 287, 322, 417, 474, 500
Pennies, as hundredths, 451, 476
Pentagons, line symmetry/line of symmetry in, 650
Performance Task, *In every chapter. For example, see:* 27, 63, 129, 191, 253, 297, 335, 401, 439, 487
Perimeter
definition of, 564
of rectangles
finding unknown measures from given, 575–580
formula for, 563–568
solving word problems involving, 581–586
Period
definition of, 4
in reading and writing multi-digit numbers, 9–14
in understanding place value, 4–8
Perpendicular lines
definition of, 606
identifying and drawing, 605–610
symbol for, 606, 609
Pints (pt)
definition of, 520
equivalent capacity in, 519–524
in mixed measures, 545, 546, 548
Place value
adding multi-digit numbers using, 39–44
adding partial sums using, 51, 52
comparing numbers using, 15–20
division using
multi-digit numbers by one-digit numbers, 235–240
by one-digit numbers, 241–246
tens, hundreds, and thousands, 199–204
two-digit numbers by one-digit numbers, 229–234

estimating sums and differences using, 33–38
groups (periods) of, 4
lining up addition problem using, 39–44
lining up subtraction problem using, 45–50
multiplication using
by multiples of 10, 75–80, 143–148
with partial products, 99–104
by tens, hundreds, or thousands, 75–80
two-digit numbers, 167–172, 179, 180
reading and writing multi-digit numbers using, 9–14
rounding multi-digit numbers using, 21–26
understanding, 3–8
Place value charts
comparing decimals on, 464, 466, 467
comparing numbers on, 15–20
definition of, 4
extending
to include hundredths, 452, 455
to include tenths, 446, 449
identifying value of digits on, 4–8
using to read and write numbers, 10–14
Platform scale, for weighing, 513
Points
definition of, 594
identifying, drawing, and naming, 593–598
Polygon mesh, 687, 688
Pounds (lb)
definition of, 514
equivalent weight in, 513–518
in mixed measures, 544–547
Precision, *Throughout. For example, see:* 57, 74, 190, 355, 462, 481, 509, 553, 625, 654
Prime numbers
definition of, 280
identifying, 279–284
Problem solving, *See* Word problems

A25

Problem Solving Plan, *Throughout. For example, see:* 58, 124, 186, 248, 396, 434, 538

Products, *See also* Multiplication
    estimating, 149–154
        choosing expression for, 149
        using compatible numbers, 149–154
        using rounding, 81–86, 149–154
    partial (*See* Partial products)
    unknown, letter representing, 123–128, 185–190

Protractor
    definition of, 624
    finding unknown angle measures with, 635–640
    measuring and adding angles with, 629–634
    measuring and drawing angles with, 623–628

Q

Quadrilaterals, classifying, 673–678

Quarts (qt)
    definition of, 520
    equivalent capacity in, 519–524
    in mixed measures, 545–548

Quotients, *See also* Division
    definition of, 218
    estimating, 205–210
    using models to find, 211–216
    partial (*See* Partial quotients)
    writing zero in, 242, 245

R

Rays
    definition of, 594
    identifying, drawing, and naming, 593–598

Reading, *Throughout. For example, see:* T-97, T-215, T-289, T-315, T-523

Reading multi-digit numbers, 9–14

Real World, *See* Modeling Real Life

Reasoning, *Throughout. For example, see:* 8, 33, 74, 145, 205, 272, 305, 364, 409, 457

Rectangles
    area of
        finding unknown measures from given, 575–580
        formula for, 569–574
        solving word problems involving, 581–586
    classifying, 674–678
    definition of, 674
    finding factor pairs using, 261–266
    line symmetry/line of symmetry in, 650, 652, 654
    perimeter of
        finding unknown measures from given, 575–580
        formula for, 563–568
        solving word problems involving, 581–586

Regrouping
    in adding multi-digit numbers, 39–44
    in dividing
        to find quotient and remainder, 212
        multi-digit numbers by one-digit numbers, 236–240
        by one-digit numbers, 241–246
        two-digit numbers by one-digit numbers, 230–234
    in multiplying
        multi-digit numbers by one-digit numbers, 111–116
        by multiples of 10, 144
        two-digit numbers, 174, 177–180
        two-digit numbers by one-digit numbers, 105–110
    in subtracting multi-digit numbers, 45–50

Remainders
    definition of, 212
    in dividing by one-digit numbers, 241–246

in dividing multi-digit numbers by
one-digit numbers, 235–240
in dividing two-digit numbers by
one-digit numbers, 229–234
using models to find, 211–216
using partial quotients with, 223–228
Repeated Reasoning, 3, 75, 143, 199, 347, 365, 617
Response to Intervention, *Throughout. For example, see:* T-1B, T-43, T-141B, T-227, T-303B, T-443B, T-493B, T-584
Review & Refresh, *In every lesson. For example, see:* 8, 38, 178, 204, 266, 310, 352, 414, 450, 500
Rhombus
classifying, 674–678
definition of, 674
lines of symmetry in, 676
pattern blocks, measuring angles using, 617, 618
Right angles
classifying quadrilaterals by, 673–678
classifying triangles by, 667–672
complementary angles of, 636–640
definition of, 600
identifying, drawing, and naming, 600–604
measuring
using degrees, 612–616
using pattern blocks, 617–622
Right triangles
classifying, 668–672
definition of, 668
Rokkaku (Japanese kite), symmetry of, 658
Rounding numbers
estimating products with, 81–86, 149–154
estimating sums and differences with, 33–38
multi-digit, using place value, 21–26
Rule(s)
definition of, 286
divisibility, 268, 274, 280, 283
number pattern, 286–290

S

Scaffolding Instruction, *In every lesson. For example, see:* T-5, T-71, T-187, T-287, T-331, T-417, T-545, T-631
Scale(s)
of drawing angles, 624
of line plot, 526
for measuring mass, 501
for measuring weight, 513
Scalene triangles
classifying, 662–666
definition of, 662
Seconds (sec)
definition of, 532
equivalent time in, 531–536
measuring elapsed time in, 538, 541, 542
in mixed measures, 545, 547
Shape patterns, creating and describing, 291–296
Shapes, two-dimensional
classifying quadrilaterals, 673–678
classifying triangles
by angles, 667–672
by sides, 661–666, 668
line symmetry in, 649–654, 676
symmetric
definition of, 656
drawing, 655–660
Show and Grow, *In every lesson. For example, see:* 4, 34, 70, 144, 200, 262, 306, 348, 410, 446
Side lengths
classifying quadrilaterals by, 673–678
classifying triangles by, 661–666, 668
finding area from, 569–574
finding perimeter from, 563–568
of rectangles, finding unknown, 575–580
Sides, parallel, 674–678
Squares
classifying, 674–678
definition of, 674
line symmetry/lines of symmetry in, 654

A27

pattern blocks, measuring angles using, 617
Standard form, of multi-digit numbers, 10–14
STEAM Performance Task, 139–140, 343–344, 559–560, 687–688
Straight angles
definition of, 600
identifying, drawing, and naming, 600–604
measuring
using degrees, 613, 616
using pattern blocks, 618
supplementary angles of, 636–640
Structure, *Throughout. For example, see:* 9, 69, 148, 211, 261, 311, 349, 411, 445, 495
Subtraction
using compensation, 51, 52, 54, 55
using "count on" strategy, 51, 52
with division, to find remainder, 218, 230–234
estimating differences in, 33–38
of fractions
using area models, 365–370
with like denominator, 371–376
in mixed numbers, 389–394
using number line, 366, 369, 372
of mixed measures, 543–548
of money amounts, 481–486
of multi-digit numbers, 45–50
strategies for, 51–56
on number line, 52
patterns in, 286–290
relationship between addition and, 390
two-step word problems solved using, 57–62
Success Criteria, *In every lesson. For example, see:* 3, 33, 69, 143, 199, 261, 305, 347, 409, 445
Sums, *See also* Addition
estimating
to check reasonableness of answer, 39–44

using rounding, 33–38
mixed numbers as, 378
partial, addition using, 51, 52
of unit fractions, writing fractions as, 353–358, 409–420
Supplementary angles
definition of, 636
finding measures of, 636–640
Symbols
degree, 612
parallel lines, 606, 609
perpendicular lines, 606, 609
Symmetric shapes
definition of, 656
drawing, 655–660
Symmetry, line
definition of, 650
identifying shapes with, 649–654, 676
Symmetry, line of
definition of, 650
for drawing symmetric shapes, 655–660
identifying, 650–654

T

Tens (10)
dividing, 199–204
multiplying, 143–148
using number line, 144, 148
using place value, 75–80, 143–148
using properties, 143–148
in number patterns, 286, 289
regrouping
in dividing by one-digit numbers, 241–246
in dividing multi-digit numbers by one-digit numbers, 236–240
in dividing two-digit numbers by one-digit numbers, 230–234
in multiplying multi-digit numbers by one-digit numbers, 111–116
in multiplying two-digit numbers, 174, 177–180

in multiplying two-digit numbers by
one-digit numbers, 105–110
rounding to nearest
estimating product by, 81–86
in estimating sums and differences,
33–38
using place value, 21–26
Tens place, 4–8
in comparing numbers, 15–20
in reading and writing numbers, 10–14
Tenths
definition of, 446
extending place value chart to include,
446, 449
money amounts in, 475–480
plotting on number line, 457
understanding, 445–450
writing as equivalent fractions and
decimals, 457–462
Tenths place, 446
Think and Grow, *In every lesson. For example,
see:* 4, 34, 70, 144, 200, 262, 306,
348, 410, 446
Think and Grow: Modeling Real Life, *In every
lesson. For example, see:* 6, 36, 72,
146, 202, 264, 308, 350, 412, 448
Thousands
dividing, 199–204
multiplying, using place value, 75–80
regrouping
in dividing by one-digit numbers,
241–246
in multiplying multi-digit numbers by
one-digit numbers, 111–116
rounding to nearest
in estimating sums and differences,
33–38
using place value, 21–26
Thousands period, 4–8
in comparing numbers, 15–20
definition of, 4
in reading and writing numbers, 10–14

Three-digit numbers, *See also* Multi-digit
numbers
multiplying by one-digit numbers,
111–116
360 degrees, 612–616
Time
elapsed
measuring on number line, 538
solving word problems involving,
537–542
mixed measures of, 544–548
units of
definitions of, 532
equivalence of, 531–536
Time intervals, 538 (*See also* Time, elapsed)
Title, for line plot, 526
Tons (T)
definition of, 514
equivalent weight in, 513–518
in mixed measures, 544, 545, 547
Trapezoids
classifying, 674–678
definition of, 674
pattern blocks, measuring angles using,
617
Triangles
acute, 668–672
classifying
by angles, 667–672
by sides, 661–666, 668
equiangular, 668–672
equilateral, 662–666, 668
isosceles, 662–666
line symmetry/line of symmetry in, 650,
651, 653
obtuse, 668–672
pattern blocks, measuring angles using,
617, 618
right, 668–672
scalene, 662–666
Two-digit numbers
dividing, by one-digit numbers, 229–234
multiplying, 173–178
using area models, 155–166, 179, 180

A29

using Associative Property, 179, 183
using Distributive Property, 161–166, 179, 183
estimating products of, 149–154
by multiples of ten, 143–148
by one-digit numbers, 105–110
using partial products, 167–172, 179, 180
using place value, 167–172, 179, 180
practicing strategies of, 179–184
using regrouping, 174, 177–180
word problems solved using, 185–190

Two-dimensional shapes
classifying quadrilaterals, 673–678
classifying triangles
by angles, 667–672
by sides, 661–666, 668
line symmetry in, 649–654, 676
symmetric
definition of, 656
drawing, 655–660

U

Unit fractions
definition of, 354
multiples of
on number line, 410, 413, 417
writing fractions as, 409–414
writing multiples of fractions as, 415–420
writing fractions as sum of, 353–358, 409–420

Unknown measures
of angles, finding, 635–640
of rectangles, finding, 575–580

Unknown numbers, letters representing
in division problems, 247–252
in multiplication problems, 123–128, 185–190

V

Vertex (vertices)
of adjacent angles, 630
definition of, 600

Vertical format
for addition problem, 39–44
for subtraction problem, 45–50

W

Weeks
definition of, 532
equivalent time in, 532–536
in mixed measures, 545, 547

Weight
customary units of, 513–518
line plot of, 527, 530
mixed measures of, 544–547

Weights, for measuring mass, 501

Which One Doesn't Belong?, *Throughout. For example, see:* 11, 119, 234, 310, 379, 405, 462, 480, 500, 616

Whole, mixed numbers and, 377–382

Whole dollars
definition of, 476
fractions/decimals and, 475–480

Whole numbers, multiplying
by fractions, 421–426
by mixed numbers, 427–432

Width
in area formula, 569–574
in perimeter formula, 563–568
of rectangles, finding unknown, 575–580

Word form, of multi-digit numbers, 10–14

Word problems, solving
with addition and subtraction, 57–62
with division, 247–252
with elapsed time, 537–542
with fractions, 395–400, 433–438
with mixed numbers, 395–400, 433–438
with multiplication
comparisons, 69–74

multi-step, 123–128, 185–190
  with two-digit numbers, 185–190
  with perimeter and area, 581–586
Writing, *Throughout. For example, see:* 5, 35, 71, 148, 222, 263, 313, 349, 426, 447

Y

Yards
  definition of, 508
  equivalent length in, 507–512
  finding area in, 570, 572, 573
  finding perimeter in, 565–568
  in mixed measures, 545, 547

Years
  definition of, 532
  equivalent time in, 532, 534, 536
  in mixed measures, 545
You Be the Teacher, *Throughout. For example, see:* 5, 41, 83, 145, 201, 266, 307, 352, 423, 456

Z

Zero (0)
  patterns, in multiplying by tens, 143–148
  writing in quotient, 242, 245

Reference Sheet

Symbols

×	multiply	°	degree(s)	$\overleftrightarrow{AB}$	line AB
÷	divide	⊥	is perpendicular to	$\overrightarrow{AB}$	ray AB
=	equals	∥	is parallel to	$\overline{AB}$	line segment AB
>	greater than	A•	point A	∠ABC	angle ABC
<	less than				

Money

¢ cent or cents
$ dollar or dollars

1 penny = 1¢ or $0.01
1 nickel = 5¢ or $0.05
1 dime = 10¢ or $0.10

1 quarter = 25¢ or $0.25
1 dollar ($) = 100¢ or $1.00

Length

Metric

1 centimeter (cm) = 10 millimeters (mm)
1 meter (m) = 100 centimeters
1 kilometer (km) = 1,000 meters

Customary

1 foot (ft) = 12 inches (in.)
1 yard (yd) = 3 feet
1 mile (mi) = 1,760 yards

Mass

1 kilogram (kg) = 1,000 grams (g)

Weight

1 pound (lb) = 16 ounces (oz)
1 ton (T) = 2,000 pounds 

Capacity

Metric

1 liter (L) = 1,000 milliliters (mL)

Customary

1 pint (pt) = 2 cups (c)
1 quart (qt) = 2 pints 
1 gallon (gal) = 4 quarts 

A33

Time

1 second

1 minute (min) = 60 seconds (sec)
1 hour (h) = 60 minutes
1 day (d) = 24 hours

1 week (wk) = 7 days
1 year (yr) = 12 months (mo)
1 year = 52 weeks

Area and Perimeter

Area of a rectangle
$A = \ell \times w$

Perimeter of a rectangle
$P = (2 \times \ell) + (2 \times w)$

length (ℓ)
width (w)

Perimeter of a polygon
$P =$ sum of lengths of sides

Angles

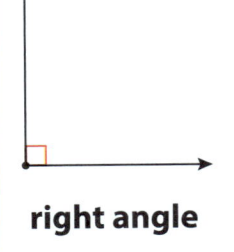

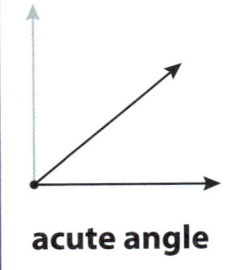

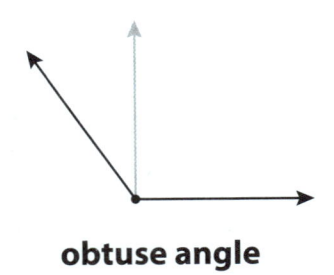

right angle

straight angle

acute angle

obtuse angle

Triangles

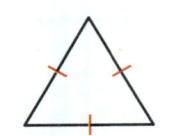

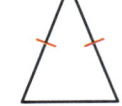

equilateral triangle

isosceles triangle

acute triangle

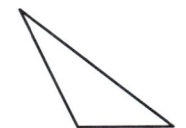

obtuse triangle

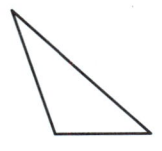

scalene triangle

right triangle

equiangular triangle

Quadrilaterals

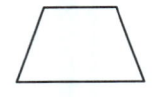

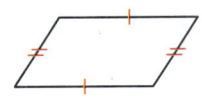

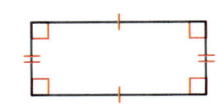

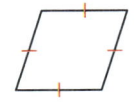

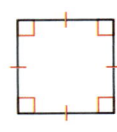

trapezoid

parallelogram

rectangle

rhombus

square

Credits

Front matter
i Brazhnykov Andriy /Shutterstock.com; **vii** Steve Debenport/E+/Getty Images

Chapter 1
1 fbxx/iStock/Getty Images Plus; **6** SSSCCC/Shutterstock.com; **8** DanielPrudek/iStock/Getty Images Plus; **11** 3DMI/Shutterstock.com; **12** Lagutkin Alexey/Shutterstock.com; **17** adventtr/iStock/Getty Images Plus; Good Life Studio/E+/Getty Images Plus; **18** Ron_Thomas/E+/Getty Images; **20** ABIDAL/iStock/Getty Images Plus; **24** Rawpixel/iStock/Getty Images Plus; **26** Maren Winter/iStock/Getty Images Plus

Chapter 2
31 SeanPavonePhoto/iStock/Getty Images Plus; **36** *top* grimgram/iStock/Getty Images Plus; Triduza Studio/Shutterstock.com; *bottom* shihina/iStock/Getty Images Plus; **38** ciud/iStock/Getty Images Plus; **41** ET-ARTWORKS/iStock/Getty Images Plus, irin717/iStock/Getty Images Plus; **42** rmbarricarte/iStock/Getty Images Plus; **44** *left* Ann W. Kosche/Shutterstock.com; *right* romakoma/Shutterstock.com; **47** efks/iStock /Getty Images Plus; **48** *left* Rawpixel/iStock/Getty Images Plus; *right* Vladimiroquai/iStock/Getty Images Plus; **50** sndr/E+/Getty Images; **54** *top* vchal/iStock/Getty Images Plus; *Exercise 14* vchal/iStock/Getty Images Plus; *bottom* vchal/iStock/Getty Images Plus; **56** SteffenHuebner/iStock/Getty Images Plus; **59** *top* Brberrys/Shutterstock.com; *bottom* anankkml/iStock/Getty Images Plus; **62** Stubblefield Photography/Shutterstock.com; Todor Rusinov/Shutterstock.com

Chapter 3
67 franckreporter/iStock/Getty Images Plus; **71** Kikkerdirk/iStock/Getty Images Plus; **72** *right* Avesun/iStock/Getty Images Plus; *left* GaryAlvis/E+/Getty Images; **74** *left* catinsyrup/iStock/Getty Images Plus; *right* mphillips007/iStock/Getty Images Plus; **77** malerapaso/E+/Getty Images; **78** *top* k_samurkas/iStock/Getty Images Plus; *bottom* Tempusfugit/iStock/Getty Images Plus; **81** Andy445/E+/Getty Images; Matt Benoit/Shutterstock.com; **83** scanrail/iStock/Getty Images Plus; **84** pictafolio/E+/Getty Images; **90** *right* luminis/iStock/Getty Images Plus; *left* neuson11/iStock/Getty Images Plus; **92** *Exercise 10* AdamParent/iStock/Getty Images Plus; *Exercise 11* cynoclub/iStock/Getty Images Plus; **96** Goddard_Photography/iStock/Getty Images Plus; *bottom* thawats/iStock/Getty Images Plus; **98** *top* CathyKeifer/iStock/Getty Images Plus; *bottom* Steve Collender/Shutterstock.com; **102** *top* lorcel/iStock/Getty Images Plus; *bottom* W6/iStock/Getty Images Plus; **104** *Exercise 12* vencavolrab/iStock/Getty Images Plus; *Exercise 13* adogslifephoto/iStock/Getty Images Plus; **108** *top* Ververidis Vasilis/Shutterstock.com; *bottom left* YinYang/iStock/Getty Images Plus; *bottom right* filrom/iStock/Getty Images Plus, Vereshchagin Dmitry/Shutterstock.com; **110** *Exercise 15* ID1974/Shutterstock.com; *Exercise 16* Nerthuz /Shutterstock.com; **113** paylessimages/iStock/Getty Images Plus; **114** *top right* leksele/iStock/Getty Images Plus; David Osborn/Shutterstock.com; *Exercise 18* itographer/E+/Getty Images; IPGGutenbergUKLtd/iStock/Getty Images Plus; **120** *top* Carso80/iStock Editorial/Getty Images Plus; *center* D3Damon/iStock/Getty Images Plus; *bottom* Iakov Filimonov/Shutterstock.com; 4kodiak/iStock Unreleased/Getty Images Plus; **122** *left* 4kodiak/iStock Unreleased/Getty Images Plus; *right* Aerotoons/DigitalVision Vectors/Getty Images; **123** Mirko Rosenau/Shutterstock.com; **125** *top left* GlobalP/iStock/Getty Images Plus; *top right* Antagain/iStock/Getty Images Plus; *bottom* roberthyrons/iStock/Getty Images Plus; **126** PeopleImages/E+/Getty Images; **128** kropic1/Shutterstock.com, ©iStockphoto.com/Chris Schmidt, ©iStockphoto.com/Jane norton; **129** *top* Maxiphoto/iStock/Getty Images Plus; *bottom* dibrova/iStock/Getty Images Plus; **130** mocoo/iStock/Getty Images Plus; ONYXprj/iStock/Getty Images Plus; Sylphe_7/iStock/Getty Images Plus; **134** magnez2/iStock Unreleased/Getty Images Plus; **136** popovaphoto/iStock/Getty Images Plus; **139** *top* TimBoers/iStock/Getty Images Plus; *bottom* Nielskliim/Shutterstock.com; **140** Wiese_Harald/iStock/Getty Images Plus

Chapter 4
141 Liufuyu/iStock/Getty Images Plus; **145** JuSun/iStock/Getty Images Plus; **146** VStock/Alamy Stock Photo; **148** Steve Collender /Shutterstock.com; **152** GlobalP/iStock/Getty Images Plus; **157** bjdlzx/iStock/Getty Images Plus; **158** *top* spanteldotru/E+/Getty Images; *bottom* AlbertoRoura/iStock Editorial/Getty Images Plus; **160** mipan/iStock/Getty Images Plus; **164** *top* bortonia/DigitalVision Vectors/Getty Images; *bottom* artisteer/iStock/Getty Images Plus, vectorloop/DigitalVision Vectors/Getty Images; **166** Thor Jorgen Udvang /Shutterstock.com; **169** GlobalP/iStock/Getty Images Plus; **170** *top* martinhosmart/iStock/Getty Images Plus; *Exercise 13* ianmcdonnell/E+/Getty Images; **176** *top* ChrisGorgio/iStock/Getty Images Plus; *Exercise 13 right* Konstantin G/Shutterstock.com; *Exercise 13 left* Triduza/iStock/Getty Images Plus; *bottom left* John_Kasawa/iStock/Getty Images Plus; **178** leezsnow/iStock/Getty Images Plus; **182** *top* Isaac74/iStock Editorial/Getty Images Plus; *bottom* breckeni/E+/Getty Images; **184** Jamesmcq24/E+/Getty Images; **185** Blade_kostas/iStock/Getty Images Plus; **186** tanyasharkeyphotography/iStock/Getty Images Plus; **187** *Exercise 4* creatOR76/Shutterstock.com; *Exercise 6* ZU_09/DigitalVision Vectors/Getty Images; **188** THEPALMER/E+/Getty Images; **189** *top* macrovector/iStock/Getty Images Plus; *bottom* RelaxFoto.de/E+/Getty Images; **190** *Exercise 6* filo/DigitalVision Vectors/Getty Images; *bottom* Aun Photographer/Shutterstock.com; **191** vladru/iStock/Getty Images Plus; **196** *Exercise 45* Believe_In_Me/iStock/Getty Images Plus; *Exercise 46* BrianAJackson/iStock/Getty Images Plus

Chapter 5
197 BanksPhotos/iStock/Getty Images Plus; **201** Grigorenko/iStock/Getty Images Plus; **202** *top* skynesher/E+/Getty Images; *bottom* quavondo/iStock/Getty Images Plus; **204** GlobalP/iStock/Getty Images Plus; **207** mar1koff/iStock/Getty Images Plus; **208** *top* Antonio V. Oquias /Shutterstock.com; *bottom* dageldog/iStock/Getty Images Plus; **210** *top* Twoellis/iStock/Getty Images Plus; *Exercise 10 left* ahirao_photo/iStock/Getty Images Plus; *Exercise 10 right* mtruchon/iStock/Getty Images Plus; **214** *top* Peter Hermus/iStock/Getty Images Plus; *bottom* kroach/iStock/Getty Images Plus; **216** duckycards/E+/Getty Images; **220** *top* Valerie Loiseleux/iStock/Getty Images Plus; *bottom* kali9/Vetta/Getty Images; **222** cynoclub/iStock/Getty Images Plus; **225** FatCamera/E+/Getty Images; **226** FatCamera/E+/Getty Images; *Exercise 18* ONYXprj/iStock/Getty Images Plus; **228** Mathisa_s/iStock/Getty Images Plus; **232** *top* Voren1/iStock/Getty Images Plus; *Exercise 18* Goce Risteski/Hemera/Getty Images; **234** *top* sarent/iStock/Getty Images Plus; *Exercise 13* tashka2000/iStock/Getty Images Plus; **237** ziherMP/iStock/Getty Images Plus; **238** *top* CTRPhotos/iStock Editorial/Getty Images Plus; *bottom* sudok1/iStock/Getty Images Plus; **240** *top* Cgissemann/iStock/Getty Images Plus; *Exercise 15* WildDoc/iStock Unreleased/Getty Images Plus; **243** humonia/iStock/Getty Images Plus; **244** McIninch/iStock/Getty Images Plus; **246** aluxum/iStock/Getty Images Plus; **247** kaanates/iStock/Getty Images Plus; **248** artplay711/iStock/Getty Images Plus; **249** *top* Monkeybusinessimages/iStock/Getty Images Plus; *bottom* Onandter_sean/iStock/Getty Images Plus; **250** AdamRadosavljevic/iStock/Getty Images Plus; **251** *left* AndreaAstes/iStock/Getty Images Plus; *right* OlegAlbinsky/iStock/Getty Images Plus; **252** *top* EdnaM/iStock/Getty Images Plus; *bottom* USO/iStock/Getty Images Plus; **253** Werner Otto/Alamy Stock Photo; **256** vladimir_n/iStock/Getty Images Plus; **258** shark_749/iStock/Getty Images Plus

Chapter 6
259 monkeybusinessimages/iStock/Getty Images Plus; **264** *top right* crossbrain66/E+/Getty Images, karandaev/iStock/Getty Images Plus; *Exercise 15* SeaHorseTwo/iStock/Getty Images Plus; **266** mipan/iStock/Getty Images Plus; **270** ivanmateev/iStock/Getty Images Plus; **272** Diana Taliun/iStock./Getty Images Plus; **276** *top* Balkonsky/Shutterstock.com; *bottom* pagadesign/E+/Getty Images; **282** *top* BruceBlock/iStock/Getty Images Plus; *bottom* brozova/iStock/Getty Images Plus; **284** kutaytanir/E+/Getty Images; **288** *top* nojustice/E+/Getty Images; *bottom* myistock88/iStock/Getty Images Plus; **290** Evgeny555/iStock/Getty Images Plus; **297** *top* leezsnow/iStock/Getty Images Plus; *center* elinedesignservices/iStock/Getty Images Plus; *bottom* stevezmina1/DigitalVision Vectors/Getty Images; **301** labsa Getty images Plus; 172969371/E+/Getty Images; ar-chi/iStock/G Plus

Chapter 7
303 kali9/iStock/Getty Images Plus; **307** stockcam/iStock/Getty Images Plus; **308** *top* Zolotaosen/iStock/Getty Images Plus; *center* repinanatoly/iStock/Getty Images Plus; *bottom* wavebreakmedia/Shutterstock.com; **310** AYImages/E+/Getty Images; **314** *top* sarahdoow/iStock/Getty Images Plus; *bottom* France68/iStock/Getty Images Plus; **316** Elnur Amikishiyev/Hemera/Getty Images Plus; stuartbur/iStock/Getty Images Plus; **320** Nikada/iStock/Getty Images Plus; **322** *left* Eyematrix/iStock/Getty Images Plus; *right* RimDream/iStock/Getty Images Plus; **325** Remus86/iStock/Getty Images Plus; **326** JoKMedia/iStock/Getty Images Plus; **328** dem10/iStock/Getty Images Plus; Dixi_/iStock/Getty Images Plus; **332** *top* suksao999/iStock/Getty Images Plus; *bottom* Krasyuk/iStock/Getty Images Plus; **334** mgkaya/iStock/Getty Images Plus; **337** AlexStar/iStock/Getty Images Plus; valery121283/iStock/Getty Images Plus; **339** pressureUA/iStock /Getty Images Plus, Rawpixel./iStock/Getty Images Plus; **340** OLEG525/iStock/Getty Images Plus; **342** kali9/E+/Getty Images; **343** GCShutter/E+/Getty Images

Chapter 8
345 FatCamera/E+/Getty Images; **350** *top* fstop123/E+/Getty Images; *bottom 15* Cheng Wei/Shutterstock.com; **356** *top* JodiJacobson/iStock/Getty Images Plus; *bottom*; Emevil/iStock/Getty Images Plus; **358** flas100/iStock/Getty Images Plus; **362** *top* Lonely__/iStock/Getty Images Plus; *bottom* etvulc/iStock/Getty Images Plus; **368** *top* pr2is/iStock/Getty Images Plus; *bottom* RonBailey/iStock/Getty Images Plus; **370** rossandgaffney/iStock/Getty Images Plus; **373** chictype/E+/Getty Images; **374** sharply_done/E+/Getty Images; **380** Quality-illustrations/iStock/Getty Images Plus; **386** Alter_photo/iStock/Getty Images Plus; **388** filo/E+/Getty Images; **392** *left* AndreaAstes/iStock/Getty Images Plus; *right* Lawrence Manning/Corbis/Getty Images; **395** *top* mazzzur/iStock/Getty Images Plus; *bottom* Taalvi/iStock/Getty Images Plus; **396** solarseven/iStock/Getty Images Plus; **397** *top* ana belen prego alvarez/iStock/Getty Images Plus; *bottom* MR1805/iStock/Getty Images Plus; **398** *top* onurdongel/iStock/Getty Images Plus; *bottom* bluestocking/E+/Getty Images; **400** PIKSEL/iStock/Getty Images Plus

Chapter 9
407 georgeclerk/E+/Getty Images; **412** *top* timsa/iStock/Getty Images Plus; *bottom* EHStock/iStock/Getty Images Plus; **414** *top* anna1311/iStock/Getty Images Plus; *bottom* AbbieImages/iStock/Getty Images Plus; **418** *top* cris180/iStock/Getty Images Plus; *bottom* StockPhotosArt/iStock/Getty Images Plus; **420** Scientifics Direct/www.ScientificsOnline.com; **424** *top* Jupiterimages/PHOTOS.com>>/Getty Images Plus; *bottom* mbbirdy/iStock Unreleased / Getty Images Plus; **430** *top* atosan/iStock/Getty Images Plus; *bottom* wickedpix/iStock/Getty Images Plus; **432** Scrambled/iStock/Getty Images Plus; **433** filo/DigitalVision Vectors/Getty Images; **434** eriksvoboda/iStock/Getty Images Plus; **435** fcafotodigital/E+/Getty Images; **436** Elenathewise/iStock/Getty Images Plus; **437** margouillatphotos/iStock/Getty Images Plus; **438** FernandoAH/E+/Getty Images; **439** mm88/iStock/Getty Images Plus

Chapter 10
443 StephanieFrey/iStock/Getty Images Plus; **444** TokenPhoto/E+/Getty Images; **447** berean/iStock/Getty Images Plus; **448** *top* Juanmonino/iStock/Getty Images Plus; *bottom* JamesBrey/E+/Getty Images; **450** *Exercise 14* master1305/iStock/Getty Images Plus; *Exercise 16* nicolamargaret/E+/Getty Images; **453** *right* sumos/iStock/Getty Images Plus; *left* OSTILL/iStock/Getty Images Plus; **454** *top* DmitriyKazitsyn/iStock/Getty Images Plus; *bottom* skegbydave/iStock/Getty Images Plus; **456** *top left* mrPliskin/E+/Getty Images; *top right* kyoshino/E+/Getty Images; *Exercise 21* Nearbirds/iStock/Getty Images Plus, dosrayitas/DigitalVision Vectors/Getty Image; **460** *top* Coprid/iStock/Getty Images Plus; *bottom* banprik/iStock/Getty Images Plus; **462** *left* Donald Miralle/DigitalVision/Getty Images; *right* FatCamera/iStock/Getty Images Plus; **466** nidwlw/iStock/Getty Images Plus **472** *top* ET1972/iStock/Getty Images Plus; *bottom* Imgorthand/E+/Getty Images; **474** *top* GlobalP/iStock/Getty Images Plus; *bottom* Chris Mattison / Alamy Stock Photo; **478** *right* Oakozhan/iStock/Getty Images Plus; *left* FuzzMartin/iStock/Getty Images Plus; **480** esanbanhao/iStock/Getty Images Plus; **483** popovaphoto/iStock/Getty Images Plus; **484** ChrisGorgio/iStock/Getty Images Plus, kbeis/DigitalVision Vectors/Getty Images; **485** ryasick/iStock/Getty Images Plus; **486** *Exercise 5* ryasick/iStock/Getty Images Plus; *Exercise 7 left* Africa Studio/Shutterstock.com; *Exercise 7 right* AlexLMX/iStock/Getty Images Plus; **487** *right* SednevaAnna/iStock/Getty Images Plus; *left* eduardrobert/iStock/Getty Images Plus; **489** DebbiSmirnoff/iStock/Getty Images Plus; **492** *left* phantq/iStock/Getty Images Plus; *right* esseffe/iStock/Getty Images Plus

Chapter 11
493 kali9/E+/Getty Images; **494** *left* PLAINVIEW/E+/Getty Images; *center* cmannphoto/iStock/Getty Images Plus; *right* wwing/iStock/Getty Images Plus; **494b** *From left to right* Ryan McVay/Photodisc; sarahdoow/iStock/Getty Images Plus; SerrNovik/iStock /Getty Images Plus; antpkr/iStock/Getty Images Plus; Canadapanda/iStock/Getty Images Plus; whitewish/iStock/Getty Images Plus; Hemera Technologies/PhotoObjects.net/Getty Images; youngID/DigitalVision Vectors/Getty Images; **498** *top* FatCamera/iStock/Getty Images Plus; *Exercise 18 left* Rvo233/iStock/Getty Images Plus; *Exercise 18 right* Onfokus/E+/Getty Images; *bottom* TonyTaylorStock/iStock Editorial/Getty Images Plus; **500** DougBennett/E+/Getty Images; **502** ikitano/iStock/Getty Images Plus; **503** Marakit_Atinat/iStock/Getty Images Plus; **504** PicturePartners/iStock/Getty Images Plus; **510** *left* amattel/iStock/Getty Images Plus; *right* cellistka/iStock/Getty Images Plus; **514** Blade_kostas/iStock/Getty Images Plus; **516** *top* Cloudtail_the_Snow_Leopard/iStock/Getty Images Plus; *bottom* DarthArt/iStock/Getty Images Plus; **518** GlobalP/iStock/Getty Images Plus; **522** Elenathewise/iStock/Getty Images Plus; **524** Photogrape/iStock/Getty Images Plus; **525** sam74100/iStock/Getty Images Plus; **527** Farinosa/iStock/Getty Images Plus; **529** Hreni/iStock/Getty Images Plus; **530** Turnervisual/iStock/Getty Images Plus; **534** JackF/iStock/Getty Images Plus; **539** Madmaxer/iStock/Getty Images Plus; **541** kali9/E+/Getty Images; **542** JonathanCohen/E+/Getty Images; **545** skaljac/Shutterstock.com; **546** jdwfoto/iStock Unreleased/Getty Images Plus; **548** Erdosain/iStock/Getty Images Plus; **549** Rich Vintage/Vetta/Getty Images; **551** mbongorus/iStock/Getty Images Plus; **552** Mastervision/Bi-Silque; **553** scubaluna/iStock/Getty Images Plus; **554** samards/iStock/Getty Images Plus; **558** Colin_Davis/iStock/Getty Images Plus; **559** Krasyuk/iStock/Getty Images Plus; **560** Pixsooz/iStock/Getty Images Plus

Chapter 12
561 dsabo/iStock/Getty Images Plus; **566** *left* finevector/iStock/Getty Images Plus; *right* ensieh1/iStock/Getty Images Plus; **568** iconogenic/iStock/Getty Images Plus; **572** *top* Eyematrix/iStock/Getty Images Plus; *bottom* GrafVishenka/iStock/Getty Images Plus; **574** denisik11/iStock/Getty Images Plus; **578** *Exercise 11* nonnie192/iStock/Getty Images Plus; *Exercise 12* Krimzoya/iStock/Getty Images Plus; **581** ismagilov/iStock/Getty Images Plus; **583** *Exercise 5* marekuliasz/iStock/Getty Images Plus; *bottom* VvoeVale/iStock/Getty Images Plus; **584** kpalimski/iStock/Getty Images Plus; **587** stocknshares/E+/Getty Images, paci77/E+/Getty Images; **590** elinedesignservices/iStock/Getty Images Plus

Chapter 13
591 JaysonPhotography/iStock/Getty Images Plus; **610** elinedesignservices/iStock/Getty Images Plus; **614** *top* Vitalliy/iStock/Getty Images Plus; *center* MicrovOne/iStock/Getty Images Plus; *bottom* szefei/iStock/Getty Images Plus; **616** Orla/iStock/Getty Images Plus; **620** *top* AnatolyM/iStock/Getty Images Plus; *center* marimo_3d/iStock/Getty Images Plus; *bottom* olgna/iStock/Getty Images Plus; **622** jonathansloane/E+/Getty Images, OLEKSANDR PEREPELYTSIA/iStock/Getty Images Plus; **634** iZonda/iStock/Getty Images Plus; **638** pialhovik/iStock/Getty Images Plus

Chapter 14
647 suebee65/iStock/Getty Images Plus; **655** adekvat/iStock/Getty Images Plus; VectorPocket/iStock/Getty Images Plus; **664** marekuliasz/iStock/Getty Images Plus; **665** *Exercise 4* Carol_Anne/iStock/Getty Images Plus; *Exercise 6* StockImages_AT/iStock/Getty Images Plus; **669** *Exercise 9* lucielang/iStock/Getty Images Plus; *Exercise 10* blue64/E+/Getty Images; *Exercise 11* Marje4/E+/Getty Images; **670** photovideostock/E+/Getty Images; **671** *Exercise 7* SlidePix/iStock/Getty Images Plus; *Exercise 8* ibooo7/iStock/Getty Images Plus; **675** *Exercise 5* Casey Botticello/iStock/Getty Images Plus; *Exercise 6* stevezmina1/DigitalVision Vectors/Getty Images; **677** antpkr/iStock/Getty Images Plus; **681** egal/iStock/Getty Images Plus; **687** Vladimir/iStock/Getty Images Plus; **688** *top* snowpeace19/iStock/Getty Images Plus; *bottom* Vladimir/iStock/Getty Images Plus

Cartoon Illustrations: MoreFrames Animation
Design Elements: oksanika/Shutterstock.com; icolourful/Shutterstock.com; Valdis Torms